Perovskite Materials, Devices and Integration

Edited by He Tian

Published in London, United Kingdom

IntechOpen

Supporting open minds since 2005

Perovskite Materials, Devices and Integration
http://dx.doi.org/10.5772/intechopen.84213
Edited by He Tian

Contributors
Jiaqi Zhang, Wubo Li, Rajan Kumar Singh, Neha Jain, Sudipta Som, Jai Singh, Ranveer Kumar, Somrita Dutta, Juan Ramón González-Velasco, Jon Ander Onrubia, Beñat Pereda-Ayo, Unai De-La-Torre, Mirela Dragan, Stanica Enache, Mihai Varlam, Konstantin Petrov, Ram Sagar Yadav, Dinesh Kumar, Monika Kanwal, Akhilesh Kumar Singh, Shyam Bahadur Rai, Sajad Dar, Ashwith Chilvery, Sharvare Palwai, Padmaja Guggilla, Kijana Wren, Antonella Glisenti, Andrea Bedon, He Tian, Xiangshun Geng, Tian-Ling Ren

First published in London, United Kingdom, 2020 by IntechOpen
IntechOpen is the global imprint of INTECHOPEN LIMITED, registered in England and Wales, registration number: 11086078, 7th floor, 10 Lower Thames Street, London, EC3R 6AF, United Kingdom
Printed in Croatia

British Library Cataloguing-in-Publication Data
A catalogue record for this book is available from the British Library

Additional hard and PDF copies can be obtained from orders@intechopen.com

Perovskite Materials, Devices and Integration
Edited by He Tian
p. cm.
Print ISBN 978-1-78985-071-0
Online ISBN 978-1-78985-072-7
eBook (PDF) ISBN 978-1-83880-848-8

For EU product safety concerns:
IN TECH d.o.o., Prolaz Marije Krucifikse Kozulić 3, 51000 Rijeka, Croatia,
info@intechopen.com or visit our website at intechopen.com.

Meet the editor

He Tian received his Ph.D. degree from the Institute of Microelectronics, Tsinghua University in 2015. He did his postdoctoral research at the University of Southern California and Yale University from 2015 to 2017. Since 2017, he joined Tsinghua University as Assistant Professor. He is currently an Associate Professor in Tsinghua University. He has been researching various novel material-based nanodevices, such as perovskite and 2D materials. Dr. Tian has published 149 papers, and his papers have been cited 3866 times, including citations from 23 academicians and 17 fellows. He has published 64 first author/corresponding author SCI papers (average impact factor 8.1), including 3 papers in Nature Communications, 20 top papers at IF>10 journal ACS Nano, Nano Letters, etc. He has hosted or participated in more than 10 scientific research projects, including NSF for Distinguished Young Scholars, NSF key projects, NSF general projects, Fok Ying-Tong Fund of the Ministry of Education, Young Elite Scientists Sponsorship Program by CAST, American AFOSR, American NSF, etc.

Contents

Preface XI

Section 1
The Overview of Applications and Prospects of Perovskite Materials 1

Chapter 1 3
Introductory Chapter: Perovskite Materials and Advanced Applications
by Xiangshun Geng, He Tian and Tian-Ling Ren

Chapter 2 11
Perovskite Materials: Recent Advancements and Challenges
by Ashwith Chilvery, Sharvare Palwai, Padmaja Guggilla, Kijana Wren and Devon Edinburgh

Chapter 3 27
Synthesis Techniques and Applications of Perovskite Materials
by Dinesh Kumar, Ram Sagar Yadav, Monika, Akhilesh Kumar Singh and Shyam Bahadur Rai

Section 2
Applications of Perovskite in Pollutant Degradation and Energy Storage 47

Chapter 4 49
Developing Functionality in Perovskites from Abatement of Pollutants to Sustainable Energy Conversion and Storage
by Andrea Bedon and Antonella Glisenti

Chapter 5 71
Perovskite-Based Formulations as Rival Platinum Catalysts for NO_x Removal in Diesel Exhaust Aftertreatment
by Jon Ander Onrubia-Calvo, Beñat Pereda-Ayo, Unai De-La-Torre and Juan Ramón González-Velasco

Chapter 6 95
Perovskite-Based Materials for Energy Applications
by Mirela Dragan, Stanica Enache, Mihai Varlam and Konstantin Petrov

Section 3
Other Applications of Perovskite 115

Chapter 7 117
Lead-Free Hybrid Perovskite Light-Harvesting Material for QD-LED Application
by Rajan Kumar Singh, Neha Jain, Sudipta Som, Somrita Dutta, Jai Singh and Ranveer Kumar

Chapter 8 137
Perovskite Materials for Resistive Random Access Memories
by Jiaqi Zhang and Wubo Li

Chapter 9 157
Osmium Containing Double Perovskite Ba_2XOsO_6 (X = Mg, Zn, Cd): Important Candidates for Half-Metallic Ferromagnetic and Spintronic Applications
by Sajad Ahmad Dar

Preface

This book provides guidance for researchers to better understand the properties of perovskite and the latest development of applications. It introduces the kinds of perovskite materials, crystal structures, physical and chemical properties, preparation methods, and recent advances regarding applications in different fields.

The authors who collaborated in this book have summarized present advances of perovskite oxides and halide perovskite in certain fields related to environment conservation and energy utilization, as well as their experience of perovskite-related research. The book contains nine chapters, organized in three sections that cover important research aspects regarding the whole perovskite system.

The first section consists of an introductory chapter prepared by the editor for the purpose of presenting a brief background on perovskite materials. It describes that perovskite is a material with an ABX_3 structure, which includes perovskite oxide and hybrid perovskite such as Ba_2XOsO_6 (X = Mg, Zn, Cd) and $Cs_2AgBiBr_6$. In this section, the crystal structures and morphologies of perovskite materials have been systematically introduced. Subsequently, we discussed the preparation methods of materials and devices. The synthesis techniques and application of perovskite oxides are discussed in detail in a separate chapter. Through the introduction of physical characteristics, it can provide help for the construction of high-performance devices. In view of the exotic properties of perovskite, the applications and development of perovskite materials in different fields including environment protection and energy utilization are illustrated.

The second section discusses environment conservation and energy utilization. Specifically, this section highlights the progress of using perovskite materials for solar cells and focuses on elucidating a few challenges of these materials in various aspects. And then, different functionalities in perovskites are reviewed from abatement of pollutants to energy conversion and storage. Take a NO_x removal as an example, this section overviews recent research on development of novel perovskite-based catalysts for NO_x removal from diesel engine exhaust gases. This section provides precise concentrations on the over-all related of perovskite materials and explores the synthesis methods and morphologies. Finally, this section reveals the significant tasks and outlooks of perovskite photocatalytic applications.

The third section presents other applications including the quantum dot LED, resistive random-access memories, and half-metallic ferromagnetic and spintronic applications. The self-issue of perovskite materials and the solution to this problem are proposed in this section, which provides a strategy for constructing high performance devices.

The editor expresses his thanks to all the participants in this book for their valuable contributions and to Ms. Nina Kalinic Babic for her assistance in finalizing the work. Acknowledgment to the IntechOpen staff members responsible for the completion of this book and other publications for free visible knowledge.

He Tian
Tsinghua University,
China

Section 1

The Overview of Applications and Prospects of Perovskite Materials

Chapter 1

Introductory Chapter: Perovskite Materials and Advanced Applications

Xiangshun Geng, He Tian and Tian-Ling Ren

1. Perovskite structure and synthesis

Perovskite is considered one of the most promising materials of the twenty-first century. In the past few decades, the perovskite has attracted broad attention and made great progress in energy storage, pollutant degradation as well as optoelectronic devices due to its superior photoelectric and catalytic properties. All materials with ABX_3 structure are collectively referred to as perovskite materials, which can be simply divided into inorganic perovskite and organic-inorganic hybrid perovskite. The tolerance factor is usually used to indicate the structure of perovskite. Each ion radius in perovskite oxide should satisfy the following equation: $t = ((r_A + r_O))/(\sqrt{2}(r_B + r_O))$, where r_O, r_A, and r_B are the radii of respective ions A, B, and oxygen elements. So far, various perovskites, such as Ba_2XOsO_6 (X = Mg, Zn, Cd), $Cs_2AgBiBr_6$, and $CH_3NH_3PbX_3$ (X = Cl, Br, I), have been synthesized and used in different fields. For example, perovskite oxides play a pivotal role in half-metallic ferromagnetic, spintronic applications, energy storage, and pollutant degradation, while the halide perovskite is used for LEDs and photodetectors. Currently, diverse preparation methods have been developed for the synthesis of perovskite with different dimensions. For instance, solid phase synthesis method and sol-gel method are used for synthesizing perovskite oxide and hydrothermal method for halide perovskite.

1.1 Solid phase synthesis

Solid phase synthesis is a traditional preparation to obtain perovskite oxides by evenly mixing two or more kinds of metal salts and pressing them into sheets. After calcining at a certain temperature, this material can be acquired by grinding calcined sample. To study its magnetism, Yuan et al. prepared the perovskite oxide $Y_{1-x}Gd_xFeO_3$ ($0 \leq x \leq 1$) with good crystal structure by solid phase method [1]. This preparation process has the advantages of simple production process and low cost, and as-prepared materials have high mechanical strength.

1.2 Sol-gel method

Organometallic compounds or inorganic metal salts as precursors are hydrolyzed or alcoholized to form sol and are finally condensed to form gel. After heat treatment, the required oxide powder is obtained. The commonly used gels include ethanol, ammonia, polyvinyl alcohol, citric acid, etc. Taguchi et al. synthesized $LaCoO_3$ with small particle size by using ethylene glycol and citric acid as gel [2]. Besides, the effects of different calcination temperatures on the properties of materials were also studied by Toro et al. [3].

1.3 Hydro-thermal synthesis

Using an aqueous solution as the reaction medium, perovskite crystals were precipitated in the reaction vessel under high temperature and pressure. Wang et al. have prepared the perovskite oxide of $LaCrO_3$, $La_{0.9}Sr_{0.1}CrO_3$, and $La_{0.8}Sr_{0.2}CrO_3$, in which the grain size is between 1 and 2 μm [4]. The crystallinity, particle size, and morphology of materials can be controlled by hydrothermal process and prepare ultra-fine, less agglomerated, and grow single-crystal spherical core-shell perovskite materials.

1.4 Vapor deposition

This technique mainly uses one or several gas phase compounds or elemental materials containing thin film elements to produce thin films by chemical reactions on the substrate surface. Liu et al. have reported the first vapor-deposited perovskite films through dual-source evaporate $PbCl_2$ and CH_3NH_3I on the FTO substrates [5]. Later, smooth and highly crystalline perovskite thin films were prepared by pulsed laser deposition [6]. Large amount of researches have revealed that the quality has a great influence on the precursor ratio control and deposition rate. Besides, chemical vapor deposition, as a general method, is also used to synthesize one-dimensional nanowires [7, 8] and two-dimensional microplatelets [9].

1.5 Solution-chemistry approaches

Solution-chemistry approaches, such as spin-coating, anti-solvent crystallization, inverse temperature crystallization, are low-cost and facile processes for preparing perovskite films and high-quality crystals. There are two strategies of one- and two-step methods about the spin-coating method. In the one-step method, perovskite precursor solution was directly applied to the substrate surface and formed perovskite film after annealing treatment. However, the main challenge of volume shrinkage has a great impact on the quality of the film. Thus, two-step

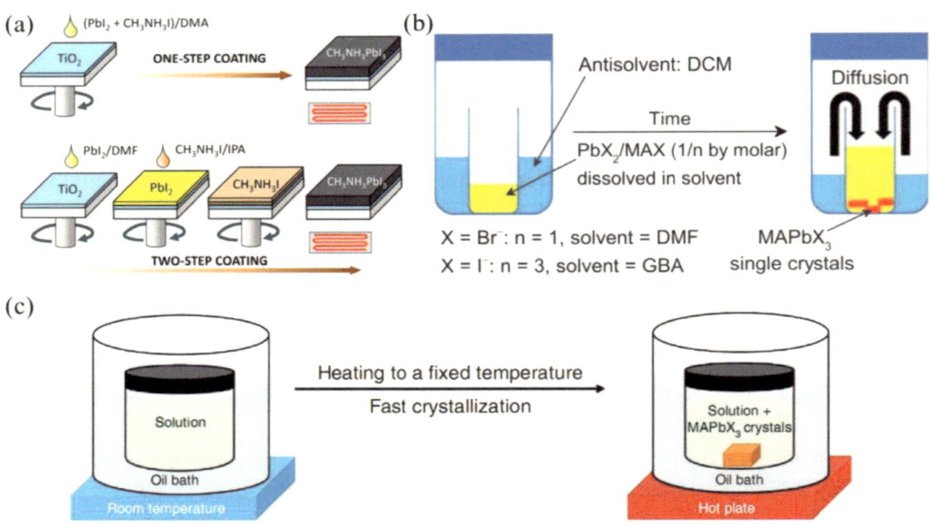

Figure 1.
(a) The illustration of spin-coating method, (b) anti-solvent crystallization, and (c) inverse temperature crystallization for preparing perovskite materials.

method is development for the preparation of uniform perovskite films, which could reduce this disadvantage to some extent. For synthesizing bulk crystals, Shi et al. proposed anti-solvent process to prepare low trap-state density and long carrier diffusion $CH_3NH_3PbBr_3$ single crystals [10]. After that, two-inch-sized perovskite $CH_3NH_3PbX_3$ (X = Cl, Br, I) crystals were prepared by inverse temperature crystallization [11]. **Figure 1** shows the schematic diagram of various solution-chemistry approaches [10, 12, 13].

2. Perovskites for devices

Perovskite has been widely used in many fields due to the great progress in material and device preparation technology. Up to date, there are numerous perovskite applications including degradation of organic pollutants, optoelectronic devices, and memory devices. Next, we will discuss perovskite applications separately.

2.1 Perovskite for catalyst

With the enhancement of environmental awareness, the construction of sustainable development society has become the current consensus. Due to the excellent catalytic properties, perovskite has a great application prospect in the degradation of organic pollutants and the acquisition of clean energy. It is well known that NO_x is one of the main causes of air pollution. Several studies were carried out to achieve NO_x-to-N_2 conversion by using perovskite-based catalysts. Furthermore, perovskite also shows exotic catalytic properties to other harmful gases, such as CO [14] and SO_2 [15], which effectively reducing pollutants in the environment. To effectively manage clean energy, previous work mainly focused on perovskite-based solid oxide fuel cell and water electrolysis.

2.2 Perovskite for optoelectronics

2.2.1 Solar cells

Perovskites are considered to be the most promising candidates for solar cells due to their excellent diffusion length (more than 1 μm), low preparation temperature, low cost, and high efficiency. In 2009, Kojima et al. first prepared perovskite solar cells [16], which is an important step for the development of solar cells. Later, Burschka et al. fabricated the solar cells by two-step continuous deposition method, and the photovoltaic conversion efficiency was increased to 15% [17]. And then, the performance was improved by changing device structure and optimizing carrier transport layer [18]. So far, the efficiency of perovskite solar cells has exceeded 22% [19]. The huge development of perovskite solar cells will provide the possibility for its commercialization.

2.2.2 Photodetectors

Photodetectors, which could convert incident light into electrical signal, are very important optoelectronic devices for optical communications, homeland security, and environmental monitoring. Many works have reported that perovskite-based photodetectors have the abilities to sense the spectra from deep-UV to visible and NIR [20, 21] and even to X-ray or γ ray [22, 23]. Efforts have been devoted to improve the device performance. For example, wide spectrum detection from

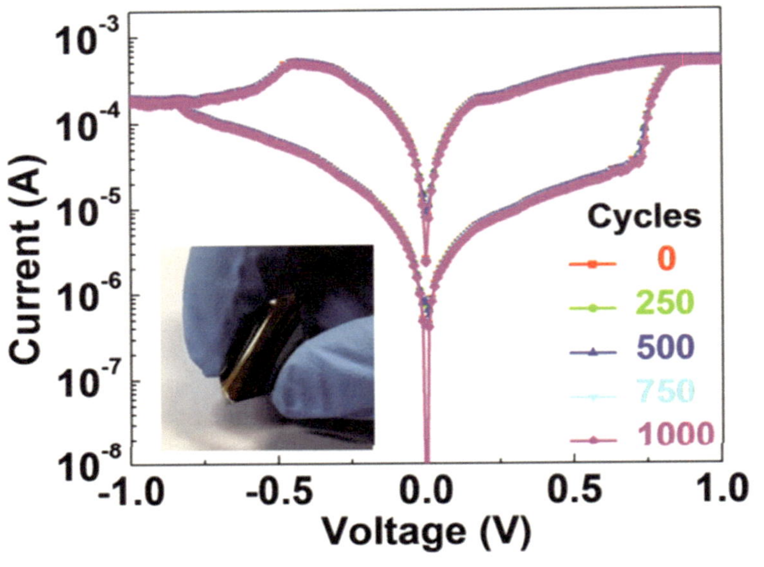

Figure 2.
Resistive switching characteristics and the photograph of the flexible device.

visible to near-infrared was realized by adjusting perovskite components [24]. The passivation of graphene makes the detectable light intensity of the device down to 1 pW/cm^2 [25]. Recently, flexible photodetectors and detector arrays have been prepared. Therefore, perovskite photodetectors are developing toward the direction of high performance and practicality.

2.3 Memory devices

Resistive memory (RRAM) is a non-volatile memory based on reversible conversion between high- and low-resistive states under the action of applied electric field. Zhang et al. were the first to demonstrate a 64-bit RRAM array utilizing perovskite oxide $Pr_{0.7}Ca_{0.3}MnO_3$ materials by a 500 nm CMOS process [26]. This RRAM array has a high/low resistance ratio larger than 1000. After that, multilayer-graphene transparent conductive electrodes were employed for flexible perovskite RRAMs [27]. Resistive switching characteristics and the photograph of the flexible device are shown in **Figure 2**. A typical structure of resistive random-access memories is a metal/insulator/metal (MIM) stack. The applied bias can adjust the operating state of the device at will, in which high resistance state (HRS) and low resistance state (LRS) can be formed. Conductive filament and uniform modes are currently recognized resistance conversion mechanisms. As the most common mechanism, ion migration and metal-insulator transition are considered to be the main cause of the filament mode. Besides, uniform resistance switching mainly includes the carrier trapping/detrapping and the ferroelectric polarization.

Author details

Xiangshun Geng, He Tian* and Tian-Ling Ren*
Institute of Microelectronics and Beijing National Research Center for Information Science and Technology (BNRist), Tsinghua University, Beijing, China

*Address all correspondence to: tianhe88@tsinghua.edu.cn
and rentl@tsinghua.edu.cn

References

[1] Yuan X, Sun Y, Xu M. Effect of Gd substitution on the structure and magnetic properties of YFeO3 ceramics. Journal of Solid State Chemistry. 2012;**196**:362-366. DOI: 10.1016/j.jssc.2012.06.042

[2] Taguchi H, Yamasaki S, Itadani A, Yosinaga M, Hirota K. CO oxidation on perovskite-type LaCoO3 synthesized using ethylene glycol and citric acid. Catalysis Communications. 2008;**9**:1913-1915. DOI: 10.1016/j.catcom.2008.03.015

[3] Del Toro R, Hernández P, Díaz Y, Brito JL. Synthesis of La0.8Sr0.2FeO3 perovskites nanocrystals by Pechini sol–gel method. Materials Letters. 2013;**107**:231-234. DOI: 10.1016/j.matlet.2013.05.139

[4] Wang S, Huang K, Zheng B, Zhang J, Feng S. Mild hydrothermal synthesis and physical property of perovskite Sr doped LaCrO3. Materials Letters. 2013;**101**:86-89. DOI: 10.1016/j.matlet.2013.03.083

[5] Liu M, Johnston MB, Snaith HJ. Efficient planar heterojunction perovskite solar cells by vapour deposition. Nature. 2013;**501**:395-398. DOI: 10.1038/nature12509

[6] Liang Y et al. Fabrication of organic-inorganic perovskite thin films for planar solar cells via pulsed laser deposition. AIP Advances. 2016;**6**:015001. DOI: 10.1063/1.4939621

[7] Chen J et al. Vapor-phase epitaxial growth of aligned nanowire networks of cesium lead halide perovskites (CsPbX3, X = Cl, Br, I). Nano Letters. 2017;**17**:460-466. DOI: 10.1021/acs.nanolett.6b04450

[8] Xing J et al. Vapor phase synthesis of organometal halide perovskite nanowires for tunable room-temperature nanolasers. Nano Letters. 2015;**15**:4571-4577. DOI: 10.1021/acs.nanolett.5b01166

[9] Wang Y et al. Chemical vapor deposition growth of single-crystalline cesium lead halide microplatelets and heterostructures for optoelectronic applications. Nano Research. 2016;**10**:1223-1233. DOI: 10.1007/s12274-016-1317-1

[10] Shi D et al. Solar cells. Low trap-state density and long carrier diffusion in organolead trihalide perovskite single crystals. Science. 2015;**347**:519-522. DOI: 10.1126/science.aaa2725

[11] Liu Y et al. Two-inch-sized perovskite CH3 NH3 PbX3 (X = Cl, Br, I) crystals: Growth and characterization. Advanced Materials. 2015;**27**:5176-5183. DOI: 10.1002/adma.201502597

[12] Im J-H, Kim H-S, Park N-G. Morphology-photovoltaic property correlation in perovskite solar cells: One-step versus two-step deposition of CH3NH3PbI3. APL Materials. 2014;**2**:081510. DOI: 10.1063/1.4891275

[13] Saidaminov MI et al. High-quality bulk hybrid perovskite single crystals within minutes by inverse temperature crystallization. Nature Communications. 2015;**6**:7586. DOI: 10.1038/ncomms8586

[14] Ciambelli P et al. AFeO3 (a=La, Nd, Sm) and LaFe1−xMgxO3 perovskites as methane combustion and CO oxidation catalysts: Structural, redox and catalytic properties. Applied Catalysis B: Environmental. 2001;**29**:239-250. DOI: 10.1016/s0926-3373(00)00215-0

[15] Trimble LE. Effect of SO2 on nitric oxide reduction over Ru-containing

perovskite catalysts. Materials Research Bulletin. 1974;**9**:1405-1412. DOI: 10.1016/0025-5408(74)90065-8

[16] Kojima A, Teshima K, Shirai Y, Miyasaka T. Organometal halide perovskites as visible-light sensitizers for photovoltaic cells. Journal of the American Chemical Society. 2009;**131**:6050-6051. DOI: 10.1021/ja809598r

[17] Burschka J et al. Sequential deposition as a route to high-performance perovskite-sensitized solar cells. Nature. 2013;**499**:316-319. DOI: 10.1038/nature12340

[18] Liu D, Kelly TL. Perovskite solar cells with a planar heterojunction structure prepared using room-temperature solution processing techniques. Nature Photonics. 2013;**8**:133-138. DOI: 10.1038/nphoton.2013.342

[19] Zhou J, Huang J. Photodetectors based on organic-inorganic hybrid lead halide perovskites. Advanced Science. 2018;**5**:1700256. DOI: 10.1002/advs.201700256

[20] Zhang Z, Zheng W, Lin R, Huang F. High-sensitive and fast response to 255 nm deep-UV light of CH3NH3PbX3 (X = Cl, Br, I) bulk crystals. Royal Society Open Science. 2018;**5**:180905. DOI: 10.1098/rsos.180905

[21] Ji C et al. Inch-size single crystal of a lead-free organic-inorganic hybrid perovskite for high-performance photodetector. Advanced Functional Materials. 2018;**28**:1705467. DOI: 10.1002/adfm.201705467

[22] Wei H et al. Dopant compensation in alloyed CH3NH3PbBr3-xClx perovskite single crystals for gamma-ray spectroscopy. Nature Materials. 2017;**16**:826-833. DOI: 10.1038/nmat4927

[23] Geng X et al. High-quality single crystal perovskite for highly sensitive X-ray detector. IEEE Electron Device Letters. 2020;**41**:256-259. DOI: 10.1109/led.2019.2960384

[24] Zhang Y, Liu Y, Li Y, Yang Z, Liu S. Perovskite CH3NH3Pb(BrxI1−x)3 single crystals with controlled composition for fine-tuned bandgap towards optimized optoelectronic applications. Journal of Materials Chemistry C. 2016;**4**:9172-9178. DOI: 10.1039/c6tc03592b

[25] Fang Y, Huang J. Resolving weak light of sub-picowatt per square centimeter by hybrid perovskite photodetectors enabled by noise reduction. Advanced Materials. 2015;**27**:2804-2810. DOI: 10.1002/adma.201500099

[26] Zhuang WW et al. Novel colossal magnetoresistive thin film nonvolatile resistance random access memory (RRAM). International Electron Devices Meeting. 2002:193-196. DOI: 10.1109/iedm.2002.1175811

[27] Jang CW, Hwang SW, Shin SH, Choi S-H. Significantly-enhanced stabilities in flexible hybrid organic-inorganic perovskite resistive random access memories by employing multilayer graphene transparent conductive electrodes. Journal of the Korean Physical Society. 2018;**73**:934-939. DOI: 10.3938/jkps.73.934

Chapter 2

Perovskite Materials: Recent Advancements and Challenges

Ashwith Chilvery, Sharvare Palwai, Padmaja Guggilla, Kijana Wren and Devon Edinburgh

Abstract

In the past decade, hybrid organic–inorganic perovskites (HOIP) have emerged as the exotic materials for the futuristic photovoltaics. The viability of low-temperature, solution-processed manufacturing and a unique blend of electronic and optical properties that has further indicated its goal towards a potential commercialization. This article clearly articulates the emergence of HOIPs and various challenges such as toxicity, hysteresis in these devices. Additionally, this chapter also makes an effort to highlight the advancements made in the perovskite materials for solar cells in the recent years, that include the Ruddlesden-Popper (RP) phase that has enabled us reach the power conversion efficiency of 28%. This phase is reportedly a lower dimensional structure than the conventional HOIP and exhibit better stability than the latter. This chapter also focuses to elucidate a few challenges of these RP phased HOIPs such as its synthesis, stoichiometry and process-ability in integrating the organic and inorganic entities.

Keywords: Ruddlesden-Popper, 2D perovskite, photovoltaics

1. Background

In the early 1990s, David Mitzi at the IBM Watson Research Centre extensively explored the structural dynamics of methylammonium lead iodide perovskite ($MAPBI_3$) material engineered for niche applications such as electronics, thin film transistors, photovoltaics (PV) and light emitting diodes [1–5]. Today, the hybrid halide perovskites are treated as the emerging semiconductor materials, as the past decade witnessed their successful application in highly-efficient photovoltaics [6]. Compounds based on this particular hybrid arrangement of organic and inorganic molecules has emerged in an unprecedented manner by the participation of over 1000 institutes across the globe and over 8000 published articles in its decade-old research efforts [7]. This article outlines a detailed overview about these materials advancements and challenges.

In the past decade, the lead (Pb)-based HOIP have attracted a great deal of interest for applications in high-performance optoelectronics and efficient photovoltaics [8–13] due to their ability to tune optical band-gaps from the violet to near infrared [14], as well as their higher absorption coefficients [15, 16] and potential for long-range electron and hole transport [17]. Furthermore, it is widely understood that perovskites exhibit a plethora of functionalities such as dielectric [18, 19], ferroelectric [5, 15, 20, 21], magnetoresistive [21], thermoelectric [22], electro-optic [23], semiconducting [24], conducting [21, 25], and superconduction [5, 26].

IntechOpen

2. Perovskite structure

A three-dimensioned (3D) perovskite structure usually adopts a stoichiometry of ABX_3 cubic structure where A is an organic cation (menthylammonium), B is a metal cation (Pb^{2+} or Sn^{2+}), and X is a halide (Cl^-, Br^- or I^-). Dieter Weber first reported the three-dimensional (3D) perovskite structure of methylammonium lead iodide ($MAPBI_3$) in 1978 [27]. There were a few other researchers who were investigating the optical properties of these materials and its feasibility for transistors and other electronics [1, 28–30]. The material was further brought to light in 2009, when Miyasaka reported a 3.81% efficient dye-sensitized solar cell with $MAPBI_3$ coated on TiO_2 surface [16]. This material was later heavily reported by various research groups across the globe due to its potential to transform photovoltaic technology and replace silicon [22, 31–34]. By natural disposition, they could potentially transform the PV technology.

The hybrid organic–inorganic perovskites (HOIP) have strong intermolecular hydrogen bonds between the amino and halide group ions, whereas the weak Vander Waals exists among the organic ions. The divalent metal ions function as the best metal cations for organic–inorganic framework. Among these combinations, those belonging to group 14 (including Sn^{2+}, and Pb^{2+}) attracted more interest due to their good optoelectronic properties and potential for low-temperature device fabrication [25, 35, 36]. The employment of least electronegative halide anions improves the perovskite structures for strong absorption over wide band gaps. The HOIPs, with a stoichiometry of perovskite crystal, have some excellent properties such as long electron–hole diffusion lengths (>100 nm) and carrier lifetimes, direct bandgap with large absorption coefficients, and low-cost solution based processing capabilities that made them best functional materials for solar cells [37, 38].

3. Perovskites for photovoltaics

HOIPs that inherit a diversified set of properties such as stronger and broader absorption, ultra-fast charge transportation, high dielectric, and swift charge recombination makes it the best suitable candidate material for PV [39, 40]. Also, its ease of fabrication, stability, cost, efficiency, and performance rivals the best thin-film PV technologies.

Perovskite materials exhibit excellent optoelectronic properties and superior device performance via two key device architectures—mesoscopic and planar—as illustrated in **Figure 1**. The ultimate goal of the proposed summer project is to study how the contact/electrode interfaces impact the mechanism of charge collection and investigate the role of ionic motion on electronic charge carriers. In the recent years, the $MAPBI_3$ has attracted a great deal of interest for applications in high-performance optoelectronic devices [41, 42] due to their ability to tune optical band-gaps from the violet to near infrared [14], as well as their higher absorption coefficients [15, 16] and potential for long-range electron and hole transport [17]. Moreover, the optical and electrical properties of HOIP single crystals are far superior to their thin film counterparts [43, 44]. So far, researchers have been able to grow perovskite structured single crystals of methylammonium lead triiodide ($CH_3NH_3PbI_3$) and methylammonium lead tribromide ($CH_3NH_3PbBr_3$) with both size and shape controlled capabilities [45, 46].

These materials potential has increased the power conversion efficiencies of solar cells from 1 to 24.2% [47, 48]. Despite these materials were gaining efficiencies but were losing on the stability and durability side due to the anomalous hysteretic effects that is typically exhibited by ferroelectric materials. So, it became critical to understand the significance of hysteresis in these materials.

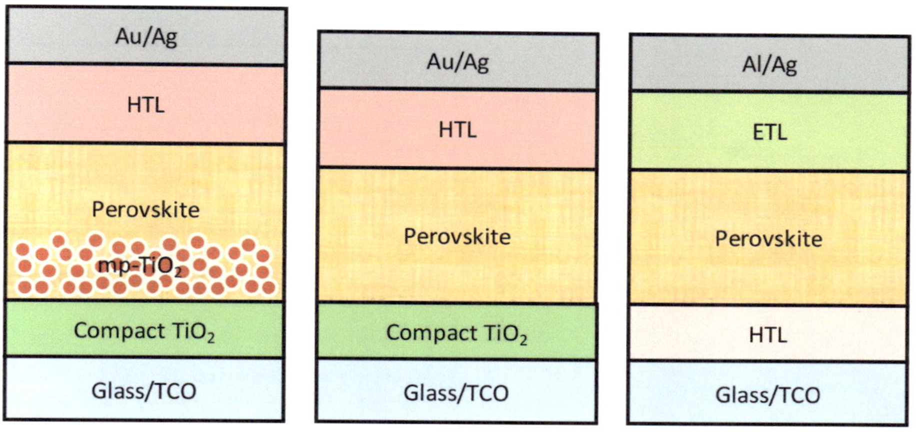

Figure 1.
Typical architectures of PSCs—mesoporous (left), conventional planar (middle) and inverted planar (right); HTL—hole transport layer, ETL—Electron transport layer, TCO—transparent conductive oxide (e. g. Indium tin oxide).

4. Hysteresis in perovskites

This unusual property of hysteresis in the J-V characteristics [49–52] of perovskites is understood to originate from various factors impacted by the various electronics parameters used to operate or test device based on these materials, such as the scan rate, sweep direction, thickness of $CH_3NH_3PbI_3$ and TiO_2. Some researchers report that the abnormal behavior is more extreme in planar structures than the mesoporous architectures [51, 53–56]. Moreover, the presence of anomalous J-V hysteresis makes it difficult to reliably quantify the device performance, which could postpone the timely commercialization of this technology. It was found that the presence of large crystal grain sizes in the perovskite layer and the use of mp-TiO_2 layer can significantly reduce the J-V hysteresis. Recently, Yang et al. reported that hysteresis highly depends on the type of perovskite material used, and device architecture (n-i-p or p-i-n) [47]. The devices with n-i-p architecture and formamidinium lead iodide ($FAPbI_3$) as the absorber material did not show any hysteresis, whereas devices with p-i-n architecture (FTO/NiO/perovskite/PCBM/LiF/Al) showed strong hysteresis.

You et al. [57] reported that their devices with ITO/NiO_x or PEDOT:PSS/$CH_3NH_3PbI_3$/ZnO or PCBM/Al (p-i-n) architecture were shown to exhibit negligible hysteresis (**Figure 2**), which is consistent with some other earlier reports [58, 59]. It was speculated that n-type PCBM layer diffuses into the perovskite layer through pin-holes and grain boundaries during processing, and mobile ions in the perovskite film may interact with PCBM, forming PCBM halide radicals. This was assumed to result in stable electrostatic properties and reduced or no hysteresis [56, 60]. On the other hand, if the PCBM is used as the bottom electron transport layer (n-i-p architecture), no such inter-diffusion occurs, and the unusual ion movement causing hysteresis cannot be suppressed. More recently, Xu et al. reported the fabrication of hysteresis-free PSCs with conventional architecture using a perovskite-PCBM hybrid solid thin film as the absorber [60]. The devices without PCBM, i.e., FTO/TiO_2/perovskite/spiro-OMeTAD/Au, exhibited hysteresis, whereas devices with a perovskite-PCBM hybrid layer showed significantly reduced or no hysteresis. Here, PCBM was shown to neutralize PbI_3^- defects, traps, and perovskite grain boundaries. It also reduced non-geminate recombination, and increased carrier lifetime. This study showed that incorporating a small amount of PCBM into the perovskite layer, or using a thin PCBM layer at the interface of TiO_2/perovskite, significantly

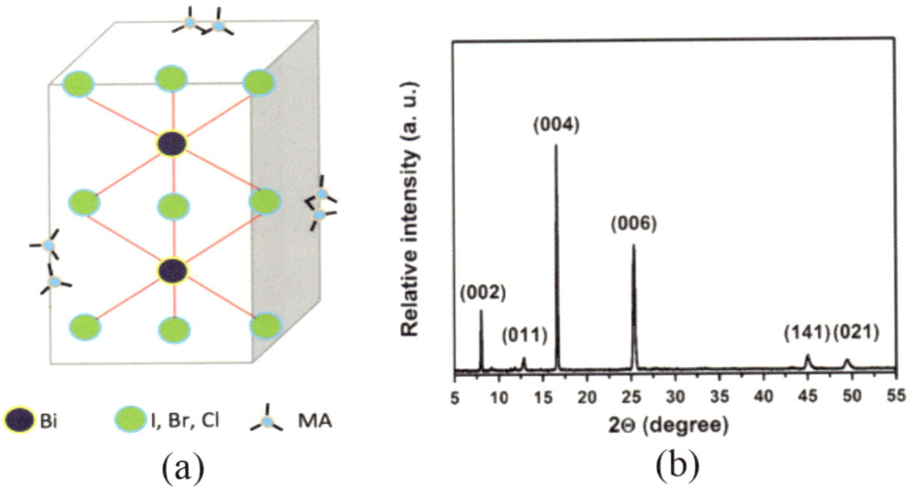

Figure 2.
(a) Schematic of MABiI perovskite structure (b) XRD studies of solution grown $(CH_3NH_3)_3Bi_2I_9$ crystal.

reduces device hysteresis, while improving the device stability. However, these observations are not consistent with the hypothesis that PCBM reduces hysteresis by preventing the diffusion of ionic defects along grain boundaries. In addition to demonstrating the role of interfacial recombination in the presence of mobile ions in a semiconductor, our study targets the demonstration of the viability of controlling this ionic migration. These measurable consequences of ion migration suggests interesting possibilities of exploiting these effects for other electronic applications (e.g. sensing) where a memory of previous operating conditions could enable novel device behavior.

The majority of investigations indicate, that an electronic alignment at the interface between the perovskite absorber and transport layer is likely to become a key ingredient for designing next-generation devices, which can overcome the current performance limits [61, 62]. Considering these various sources potentially contributing to hysteresis, further investigation into device architectures and materials design is required to understand the dominant sources of this unusual behavior.

Further, the architecture of PSCs is also found to affect the level of J-V hysteresis. Planar PSCs on cp-TiO_2 often show more severe hysteresis than mesoporous PSCs based on a mp-TiO_2 scaffold [50]. For mesoporous/planar hybrid PSCs with a perovskite capping layer on top of mp-TiO_2, the grain size and thickness of the capping layer influences the J-V response. As ionic migration was identified as a possible cause of PV current–voltage hysteresis in hybrid perovskite solar cells, it was concluded that the quality of materials and crystal growth kinetics are vital in improving electronic, optoelectronic and PV applications for these classes of hybrid organic–inorganic semiconductors. Apart from hysteresis, another drawback that was hindering these devices into the commercial markets was the toxicity of lead.

5. Bismuth-based perovskites

The HOIP class of materials have displayed promising results for advanced and efficient photovoltaics. However, some of the urgent issues associated with these materials such as lead toxicity, poor stability, and atmospheric impacts are yet to be resolved. So, the focus of research shifted to replace toxic lead with other cations such as Sn^{2+}, Ge^{2+}, Mg^{2+}, Mn^{2+}, Ni^{2+}, and Co^{2+} [63–65]. Many proposed the replacement of Pb by tin (Sn) in order to render perovskite light absorbers less toxic or even nontoxic [66, 67]. Yet, compared to their Pb-based counterparts, Sn-based perovskites so far show inferior device performance, and they are even more sensitive to degradation because of self-oxidation [24]. In addition, Sn is also listed as a harmful chemical, raising concerns regarding its suitability as a more environment friendly alternative to lead in perovskite solar cells [68]. The most obvious route to replacing Pb in this compound is via substitution of other group-14 elements, such as Sn which tend to undergo oxidation, for example, from Sn^{2+} to Sn^{4+}, leading to a rapid degradation of the corresponding halide perovskites [69].

Germanium (Ge) has received relatively little attention as a replacement to lead, in part due to its strong lone pair effect, which acts to produce lower dimensionality structures with poor conductivities [70]. The series of perovskite-structured $CsGeX_3$ compounds, where X is chlorine (Cl), bromine (Br), and Iodine (I) have been known for some time for their nonlinear optical properties [71, 72]. These compounds are stable up to 150°C, and have bandgaps correlated with the A-site cation size [73]. However, the relative stability of the +4 oxidation states decrease and the relative stability of the +2 oxidation states increase down the group IV elements. As this effect is even more prominent for the case of Ge^{2+}, it is presumptive why no successful Ge-based hybrid perovskites have been developed.

These reports indicate that neither Pb nor Sn nor Ge can steer the scientific community to a roadmap for stable and non-toxic optoelectronic devices. Thus, more out-of-the-box approaches such as blending and splitting the anions and cations is required to stimulate the research in these niche areas. Fundamentally, many of the excellent properties seen in hybrid perovskites are thought to be derived from the 6 s [2] electronic configuration of lead, a configuration seen in a range of post-transition metal compounds [74]. Typically, materials containing post-transition metals with an *ns* [2] electronic configuration (*i.e.* an N-2 oxidation state) possess large dielectric constant, small effective masses, a valence band maximum composed of antibonding states, and high levels of band dispersion properties due to their soft polarizability—leading to high Born effective charges and large spin–orbit effects, which act to increase the bandwidth of the conduction band.

One promising alternative is to incorporate **Bismuth (Bi)** into the perovskite structure. Its size is similar to lead and the impact of steric effects on the optoelectronic properties of the material. Bismuth(III) (Bi^{3+}) compounds are attractive reagents and catalysts in organic synthesis because of their low toxicity, low cost and ease of handling. Preliminary investigations on organic–inorganic methylammonium bismuth iodide (MABiI) shows its advantages of non-toxicity, stability in ambient air and low-temperature solution-processibility, which may address both the toxicity issue of lead-based perovskites and instability concern of tin-based ones [75]. Typically for methylammonium bismuth halides, a Goldschmidt tolerance factor (t) of 0.85–0.87 (Bi^{3+} = 1.03A°) is observed for a suitable perovskite structure and to further compliment it an octahedral factor (μ) Of 0.46 justifies a stable octahedron perovskite structure [76]. Another study reports that compounds formed with 6*p* block cations show a significantly larger contribution of spin-orbit coupling in the conduction band, *vs*. those formed from the 5*p* block. This implies that the Bi^{3+} compounds show a much stronger spin-orbit contribution than the In, Sn, and Sb compounds. This further reiterates that Bi based compounds are the most suitable for achieving similar shallow vacancy levels as observed in $MAPBI_3$ [77].

Organic–inorganic bismuth iodide based hybrid materials have attracted interest because of their potential semiconducting character, rich structural diversity and interesting electronic and optical properties, for which the 6 s [2] lone pair of the Bi^{3+} plays an important role in the various material properties [78, 79]. In early 1990's, researchers conducted extensive studies on the phase transitions of various

MBH primarily to investigate its application as scintillators in radiation detectors [80, 81]. Recently, several research groups reported methylammonium bismuth iodides ($(CH_3NH_3)_3Bi_2I_9$) as a stable and lead-free light absorber for PV applications with an efficiency of less than 1% [75, 82–85]. Our basic studies on Bi based single crystals of $(CH_3NH_3)_3Bi_2I_9$ were successful and in agreement with the literature (publication pending). **Figure 2a** illustrates a typical schematic of MABiI structure comprising of bioctahedral $(Bi_2I_9)^{3-}$ clusters that are surrounded by methylammonium cations. The MABiI crystal was solution grown at ambient room temperatures with good hexagonal crystalline structural phase and X-ray diffraction as shown in **Figure 2b** is in agreement with the literature [75, 83]. The red single crystals suitable for single crystal X-ray diffraction were grown using a layer crystallization technique by separating the two starting solutions of methyl ammonium iodide and bismuth(III)iodide with dichloromethane. In our preliminary study, all of the peaks in the XRD pattern can be indexed to a hexagonal structure, with no traces of any impurities. There are reports on XRD patterns of thin films, prepared via spin coating techniques recorded in Bragg–Brentano geometry and show good agreement when compared with the pattern calculated from single crystal data demonstrating a high degree of phase purity in the new material [79, 82]. A recent study on $Cs_3Bi_2I_9$ perovskite crystal, revealed it had a hexagonal structure and its space group is $P6_3/mmc(194)$ which also has a monoclinic phase with space group C12/c1 [15] [83]. However, there are no reports for Bi based single crystal nanostructures.

As the research was plateauing on the lead alternatives and the observed hysteresis, the scientific community has identified a new breakthrough for these materials i.e., a lower dimensional perovskite- halide based two dimensioned organic–inorganic perovskite.

Recently, Benabdallah et al., have reported on the advantages of MABiI over lead-based perovskites [86]. The first principle calculations based on DFT have revealed that the Bi-based perovskites could be engineered to better bandgaps and have superior absorption coefficients than its lead counterparts. Separately, Sanders et al., have also reported on a novel chemical vapor deposition method to fabricate Bi-based perovskite solar cells that could potentially pave way for large scale and cost-effective commercial markets [87].

6. Transition from 3D to 2D crystal structures

Lately, these materials have unlocked a new stoichiometry-two-dimensional (2D), that features improved stability against atmosphere than the 3D materials. This process of lowering the dimensionality is gaining momentum and potentially serve the needs of specific applications such as solar cells, light emitting diodes (LEDs) and other optoelectronic devices. Although, there are various techniques to obtain this exotic and new 2D material, there has been significant interest for a specific kind of perovskites, namely Ruddlesden-Popper (RP) which was discovered in 1957 [88]. These perovskites have Vander Waals crystal structure carrying a general chemical formula of $L_2A_{n-1}B_nX_{3n+1}$, where L is a large molecular chained cation, A is a regular cation, B is an inorganic metal cation, X is the halide anion and n = 1, 2, etc., this arrangement of alternating organic–inorganic layers develops an ordered and ideal 2D quantum well. Additionally, these 2D structured quantum wells play a vital role in neutralizing the higher exciton binding energies formed due to contrasting dielectric constants. A simple structure is illustrated in **Figure 3**.

The insulating nature of the organic spacer cations creates a competitive advantage of greater atmospheric stability for the 2D HOIPs when compared to their 3D counterparts. It was rightly indicated by Mao et al. that the discovery of these 2D

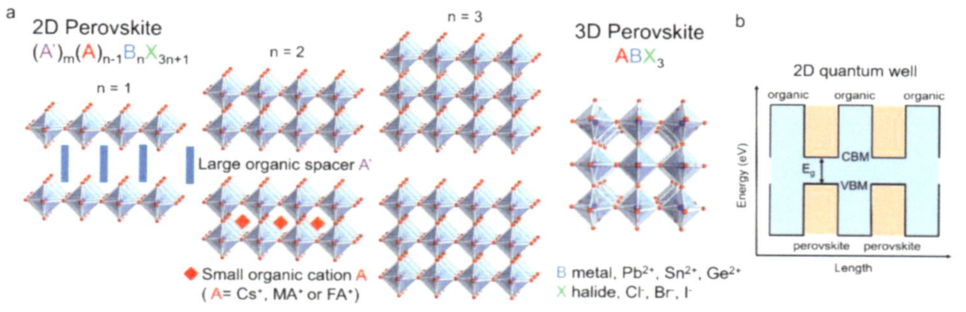

Figure 3.
(a) An illustration of the evolution from 2D perovskite to 3D perovskite with key components. (b) Representation of a 2D quantum well structure. "Reprinted with permission from Mao et al. [6]. Copyright (2019) American Chemical Society."

	L	A	B	X	n
Notation	**Bulky organic cation**	**Small organic cation**	**Inorganic cation**	**Halide ion**	**integer**
Features	Must contain functional cations to interact with inorganic anions	Insulating nature	Divalent transition metals	Key for photoconductivity	Indicates number of metal halide layers held between the two cation layers
Most commonly used materials	Butyl ammonium (n-BA),2-phenylethylammonium (PEA), ethylenediammonium (EDA)	Cs^+, MA^+	Pb^{2+}, Sn^{2+}	Cl^-, Br^-, I^-	1,2,3...
Impact on the device	Its hydrophobic nature exhibits superior ambient stability	Longer diffusion lengths for electrons and holes	Organic–inorganic layers generate quantum wells	Makes the charge transport anisotropic	Critical factor to tune the layer thickness

Table 1.
Key requirements of 2D HOIPs.

HOIPs could possibly be a solution to stabilize a few metastable phases that are found in the form of perovskite nanocrystals. As Spanopoulos et al., has rightly observed that the room-temperature PXRD measurements of these 2D perovskites have indicated that there are no residual low-angle peaks (<13° 2θ) which would be an indication of coprecipitation of more than one inorganic layer in each case [89]. Although there has been significant progress in the [90], (110) and (111)-oriented perovskites, there has been undivided attention for the (100)-oriented HOIPs due to their dimensional variability. Furthermore, this particular orientation is the most favored geometry to tune the layer thickness. A few researchers have expanded the family of 2D RP perovskites by increasing the organic spacer molecules (n = 1–5) and concluded that for n values higher than 5 does not exhibit sufficient thermodynamic

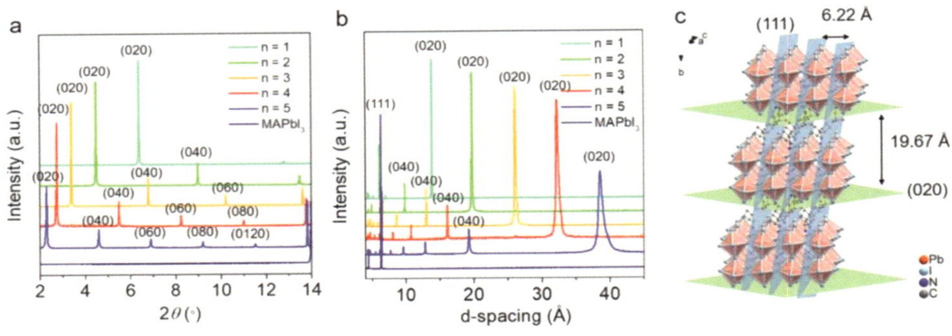

Figure 4.
(a) High-resolution PXRD of $(BA)_2(MA)_{n-1}Pb_nI_{3n+1}$ (n = 1–5). (b) PXRD plotted in d-spacing. (c) Structure of $(BA)_2(MA)Pb_2I_7$ with the illustration of respective diffraction planes. "Reprinted with permission from Mao et al. [6]. Copyright (2019) American Chemical Society."

stability [6]. Separately, other study indicated that these materials have demonstrated unique optical characteristics with high air, heat and light stability especially for n = 4. Additionally, other discoveries such as the stability increases with increasing layer thickness and the need for crystalline substrates for improved orientation and long-term air stability [89]. So, a plethora of permutations are to be considered to build an effective, stable and functional 2D HOIP. **Table 1** delineates key features about individual elements of these HOIPs.

As a result of its ability to dimensionally customize the n-layer, the [90] oriented 2D HOIPs are paving way for a precise tailoring of the quantum well electronic structures. A simple X-ray diffraction analysis as shown in **Figure 4** would help us track the number of perovskite layers. This particular orientation exhibits a level of resistance to the thermodynamic phase. For example, this phase is more evident in the 2D perovskites at n = 2 versus a non-perovskitic phase in a 3D perovskite [6]. This further reinforces the belief that these 2D materials can further emerge with unexplored photophysics and optoelectronic properties.

7. 2D perovskites for photovoltaics

Despite the fact that, these Vander Waal layered 2D structure of RP perovskites has complicated the growth kinetics of thin 2D perovskites, yet the mechanical exfoliation process of inserting the long organic cation into a 3D perovskite structure was investigated by various research groups [6, 91, 92]. A few other groups have explored other methods such as colloidal synthesis [6, 74, 93], chemical vapor deposition (CVD) [94] and solution-based methods to obtain these 2D perovskites.

Another key feature that distinguishes 2D perovskites with their 3D counterparts is its superior susceptibility towards atmospheric moisture. By far, this is believed to be the single largest factor holding its foray into commercial markets. Apart from photoinduced oxidative degradation, 3D perovskites react with water, leading to collapse of the structure and to formation of either a 0D hydrated phase or to irreversible decomposition into the precursors, for example, MAPBI3 starts to decompose at a relative humidity of 55%, which induces a color change from dark brown to yellow owing to conversion into a lead iodide [95]. Separately, other factors such as localized UV heating and light induced ion movement has also accelerated the perovskite degradation [90, 96–98].

Although, there are various synthesis methods for these 2D perovskite materials such as CVD, liquid and mechanical exfoliation, solution and colloidal based methods that have shown promise. But, there seem to be a critical need for controlled growth

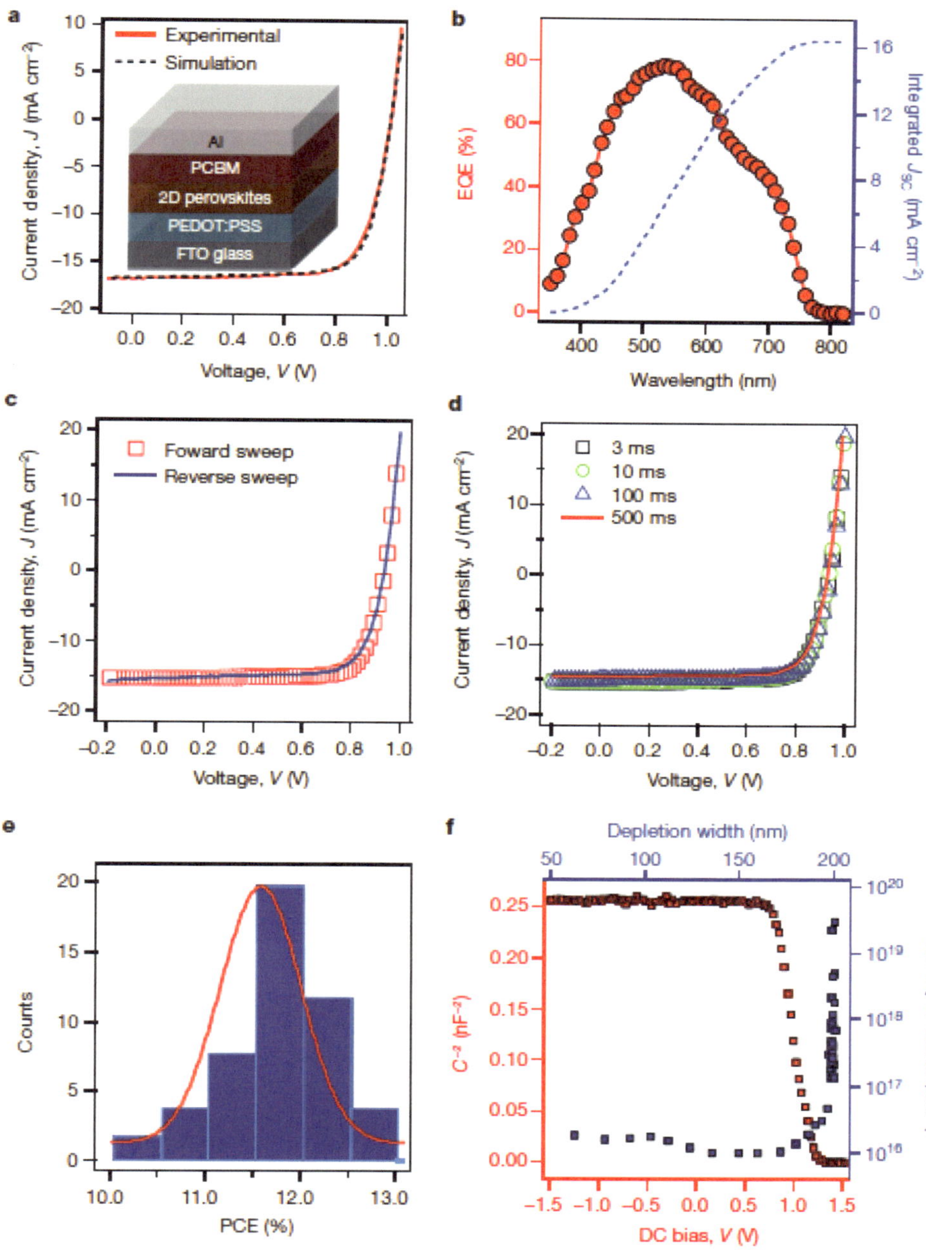

Figure 5.
Solar cell architecture and characterization. (a) Experimental (red line) and simulated (black dashed line) current-density–voltage (J–V) curves under an AM.1.5G solar simulator for planar devices. (b) External quantum efficiency (EQE; red circles and line) and integrated short- circuit current density (JSC; blue dashed line) as a function of wavelength. (c) and (d) J–V curves for hysteresis tests under AM1.5G illumination measured with the voltage scanned in opposite directions and with varying voltage delay times. (e), histogram of $(BA)_2(MA)_3Pb_4I_{13}$ device power conversion efficiency (PCE) over 50 measured devices, fitted with a Gaussian distribution (red line). (f), C–V curves for a typical device detected by a small-amplitude a.c. field (peak-to-peak voltage V_{PP} = 20 mV) at an AC frequency of 100 kHz. Reprinted with permission from Tsai et al. [99].

monitoring especially the thickness and lateral crystal sizes is still a challenge [92]. So, new synthesis methods that warrant controlled thickness and large lateral sizes are highly preferred to institute a seamless pathway towards commercial markets.

Tsai et al., have reported a uniform and thin layered films for planar PV applications as shown in **Figure 5**. These hot-casted films of 2D $(BA)_2(MA)_3Pb_4I_{13}$ perovskite was reproducible and highly efficient with 12.51% with an open circuit voltage of 1.01 V, short circuit current density of 16.76mAcm^{-2} and a fill factor of 74.13% [99]. Also, from **Figure 5c** it can be noticed that the J–V curves for hysteresis tests under AM1.5G illumination measured with the voltage scanned in opposite directions were observed to be no effects.

8. Conclusions

With the least scope for degradation and hysteresis, these 2D HOIPs have rejuvenated the scientific community at large and paved the way for a larger role in the commercial markets. Contrarily, as many consider that the dimensional transition of perovskite materials from 3D to 2D clearly outlines the scope of these materials and applications that it could create, there are equally concerning challenges such as complex growth mechanism, low thermal conductivity and high exciton binding energies. So, a reproducible and scalable fabrication mechanism is highly required to lock its potential. Also, the emergence of 2D nanostructures of HOIPs could be leading the research and evaluate the potential of these materials to its fullest. So, a methodology that can create a scalable and efficient PV devices with longer stability and improved functionalities could change the dynamics of these materials.

Acknowledgements

This work is partly supported by Louisiana NASA SAR project and NSF Awards #1546965.

Author details

Ashwith Chilvery[1]*, Sharvare Palwai[2], Padmaja Guggilla[2], Kijana Wren[1] and Devon Edinburgh[1]

1 Department of Physics, Xavier University of Louisiana, New Orleans, Louisiana, United States

2 Department of Physics, Alabama A&M University, Huntsville, Alabama, United States

*Address all correspondence to: achilver@xula.edu

References

[1] Kagan C, Mitzi D, Dimitrakopoulos C. Organic-inorganic hybrid materials as semiconducting channels in thin-film field-effect transistors. Science. 1999;**286**:945-947

[2] Mitzi DB, Prikas MT, Chondroudis K. Thin film deposition of organic-inorganic hybrid materials using a single source thermal ablation technique. Chemistry of Materials. 1999;**11**:542-544

[3] Mitzi DB. Templating and structural engineering in organic–inorganic perovskites. Dalton Transactions. 2001;**1**:1-12

[4] Mitzi DB, Chondroudis K, Kagan CR. Organic-inorganic electronics. IBM Journal of Research and Development. 2001;**45**:29-45

[5] Mitzi DB. Synthesis, Structure, and Properties of Organic-Inorganic Perovskites and Related Materials. West Sussex, England: John Wiley & Sons; 1999

[6] Mao L, Stoumpos CC, Kanatzidis MG. Two-dimensional hybrid halide perovskites: Principles and promises. Journal of the American Chemical Society. 2019;**141**:1171-1190

[7] Kumar Jena A, Kulkarni A, Miyasaka T. Halide perovskite photovoltaics: Background, status, and future prospects. Chemical Reviews. 2019;**119**:3036-3103

[8] Huang J, Shao Y, Dong Q. Organometal Trihalide perovskite single crystals: A next wave of materials for 25% efficiency photovoltaics and applications beyond? Journal of Physical Chemistry Letters. 2015;**6**:3218-3227

[9] Bade SGR et al. Fully printed halide perovskite light-emitting diodes with silver nanowire electrodes. ACS Nano. 2016;**10**:1795-1801

[10] Pasquarelli RM, Ginley DS, O'Hayre R. Solution processing of transparent conductors: From flask to film. Chemical Society Reviews. 2011;**40**:5406

[11] Fan J, Jia B, Gu M. Perovskite-based low-cost and high-efficiency hybrid halide solar cells. Photonics Research. 2014;**2**:111

[12] Chilvery AK et al. Perovskites: Transforming photovoltaics, a mini-review. Journal of Photonics for Energy. 2015;**5**:057402

[13] Chilvery AK, Guggilla P, Batra AK, Gaikwad DD, Currie JR. Efficient planar perovskite solar cell by spray and brush solution-processing methods. Journal of Photonics for Energy. 2015;**5**: 053093

[14] Knutson JL, Martin JD, Mitzi DB. Tuning the band gap in hybrid tin iodide perovskite semiconductors using structural templating. Inorganic Chemistry. 2005;**44**:4699-4705

[15] Frost JM et al. Atomistic origins of high-performance in hybrid halide perovskite solar cells. Nano Letters. 2014;**14**:2584-2590

[16] Kojima A, Teshima K, Shirai Y, Miyasaka T. Organometal halide perovskites as visible-light sensitizers for photovoltaic cells. Journal of the American Chemical Society. 2009;**131**:6050-6051

[17] Xing G et al. Long-range balanced electron- and hole-transport lengths in organic-inorganic CH3NH3PbI3. Science. 2013;**342**:344-347

[18] Juarez-Perez EJ et al. Photoinduced Giant dielectric constant in Lead halide Perovskite solar cells. Journal of Physical Chemistry Letters. 2014;5:2390-2394

[19] Batra AK et al. Simulation of energy harvesting from roads via pyroelectricity. Journal of Photonics for Energy. 2011;**1**:014001

[20] Bokov AA, Ye ZG. Recent progress in relaxor ferroelectrics with perovskite structure. Journal of Materials Science. 2006;**41**:31-52

[21] Loi MA, Hummelen JC. Hybrid solar cells: Perovskites under the sun. Nature Materials. 2013;**12**:1087-1089

[22] Takahashi Y et al. Charge transport in tin-iodide perovskite: Origin of high conductivity. Dalton Transactions. 2011;**40**:5563-5568

[23] Bhalla A, Guo R, Roy R. The Perovskite structure- a review of its role in ceramic science and technology. Materials Research Innovations. 2000;**4**:3-26

[24] Stoumpos CC, Malliakas CD, Kanatzidis MG. Semiconducting tin and lead iodide perovskites with organic cations: Phase transitions, high mobilities, and near-infrared photoluminescent properties. Inorganic Chemistry. 2013;**52**:9019-9038

[25] Mitzi DB, Feild CA, Harrison WT, Guloy AM. Conducting tin halides with a layered organic-based perovskite structure. Nature. 1994;**369**:467-469

[26] Kim H-S, Im SH, Park N-G. Organolead halide Perovskite: New horizons in solar cell research. Journal of Physical Chemistry C. 2014;**118**:5615-5625

[27] Weber D. CH3NH3SnBrxI3–x (x = 0-3), a Sn(II)-system with the cubic perovskite structure. Zeitschrift für Naturforschung Part B. 1978;**33**:1443-1445

[28] Mashiyama H, Kurihara Y. Disordered cubic Perovskite structure of CH 3 NH 3 PbX 3 (X = Cl, Br, I). Journal of the Korean Physical Society. 1998;**32**:156-158

[29] Kitazawa N, Watanabe Y, Nakamura Y. Optical properties of CH 3 NH 3 PbX 3 (X = halogen) and their mixed-halide crystals. Journal of Materials Science. 2002;**37**:3585-3587

[30] Condeles JF, Lofrano RCZ, Rosolen JM, Mulato M. Stoichiometry, surface and structural characterization of lead iodide thin films. Brazilian Journal of Physics. 2006;**36**:320-323

[31] Snaith HJ. Estimating the maximum attainable efficiency in dye-sensitized solar cells. Advanced Functional Materials. 2010;**20**:13-19

[32] Snaith H, Gratzel M. Enhanced charge mobility in a molecular hole transporter via addition of redox inactive ionic dopant: Implication to dye-sensitized solar cells. Applied Physics Letters. 2006;**89**:262114

[33] Im J-H, Lee C-R, Lee J-W, Park S-W, Park N-G. 6.5% efficient Perovskite quantum-dot-sensitized solar cell. Nanoscale. 2011;**3**:4088-4093

[34] Kim H-S et al. Lead iodide perovskite sensitized all-solid-state submicron thin film mesoscopic solar cell with efficiency exceeding 9%. Scientific Reports. 2012;**2**:591

[35] Koutselas IB, Ducasse L, Papavassiliou GC. Electronic properties of three- and low-dimensional semiconducting materials with Pb halide and Sn halide units. Journal of Physics. Condensed Matter. 1996;**8**:1217-1227

[36] Gao P, Grätzel M, Nazeeruddin MK. Organohalide lead perovskites for photovoltaic applications. Energy & Environmental Science. 2014;**7**:2448-2463

[37] Stranks SD et al. Electron-hole diffusion lengths exceeding

1 micrometer in an Organometal Trihalide Perovskite absorber. Science. 2013;**342**:341-344

[38] Ponseca CS et al. Organometal halide perovskite solar cell materials rationalized: Ultrafast charge generation, high and microsecond-long balanced mobilities, and slow recombination. Journal of the American Chemical Society. 2014;**136**:5189-5192

[39] De Wolf S et al. Organometallic halide Perovskites: Sharp optical absorption edge and its relation to photovoltaic performance. Journal of Physical Chemistry Letters. 2014;**5**:1035-1039

[40] Kazim S, Nazeeruddin MK, Grätzel M, Ahmad S. Perovskite as light harvester: A game changer in photovoltaics. Angewandte Chemie (International Ed. in English). 2014;**53**:2812-2824

[41] Zhao Y, Zhu K. Organic–inorganic hybrid lead halide perovskites for optoelectronic and electronic applications. Chemical Society Reviews. 2016;**45**:655-689

[42] Yongping F et al. Solution growth of single crystal Methylammonium Lead halide Perovskite nanostructures for optoelectronic and photovoltaic applications. Journal of the American Chemical Society. 2015;**137**:5810-5818

[43] Shi D et al. Low trap-state density and long carrier diffusion in organolead trihalide perovskite single crystals. Science. 2015;**347**:519-522

[44] Dong Q et al. Electron-hole diffusion lengths > 175 um in solution-grown CH 3 NH 3 PbI 3 single crystals. Science. 2015;**347**:967-970

[45] Saidaminov MI et al. High-quality bulk hybrid perovskite single crystals within minutes by inverse temperature crystallization. Nature Communications. 2015;**6**:7586

[46] Maculan G et al. CH3NH3PbCl3 single crystals: Inverse temperature crystallization and visible-blind UV-Photodetector. Journal of Physical Chemistry Letters. 2015;**6**:3781-3786

[47] Yang WS et al. High-performance photovoltaic perovskite layers fabricated through intramolecular exchange. Science. 2015;**348**:1234-1237

[48] Heo JH et al. Efficient inorganic-organic hybrid heterojunction solar cells containing perovskite compound and polymeric hole conductors. Nature Photonics. 2013;**7**:486-491

[49] Calado P et al. Evidence for ion migration in hybrid perovskite solar cells with minimal hysteresis. Nature Communications. 2016;**7**:1-40

[50] Chen B, Yang M, Priya S, Zhu K. Origin of J-V hysteresis in Perovskite solar cells. Journal of Physical Chemistry Letters. 2016;**7**:905-917

[51] Unger EL et al. Hysteresis and transient behavior in current-voltage measurements of hybrid-perovskite absorber solar cells. Energy & Environmental Science. 2014;**7**:3690-3698

[52] Miyano K, Yanagida M, Tripathi N, Shirai Y. Hysteresis, stability, and ion migration in Lead halide Perovskite Photovoltaics. Journal of Physical Chemistry Letters. 2016;**7**:2240-2245

[53] Frost JM, Butler KT, Walsh A. Molecular ferroelectric contributions to anomalous hysteresis in hybrid perovskite solar cells. APL Materials. 2014;**2**:081506

[54] Snaith HJ et al. Anomalous hysteresis in Perovskite solar cells. Journal of Physical Chemistry Letters. 2014;**5**:1511-1515

[55] Kim HS, Park N-G. Parameters affecting I-V hysteresis of CH 3 NH 3

PbI_3 Perovskite solar cells: Effects of Perovskite crystal size and Mesoporous TiO_2 layer. Journal of Physical Chemistry Letters. 2014;5:2927-2934

[56] Shao Y, Xiao Z, Cheng B, Yongbo Y, Jinsong H. Origin and elimination of photocurrent hysteresis by fullerene passivation in $CH_3NH_3PbI_3$ planar heterojunction solar cells. Nature Communications. 2014;**5**

[57] You J et al. Improved air stability of perovskite solar cells via solution-processed metal oxide transport layers. Nature Nanotechnology. 2015;**11**:75-81. DOI: 10.1038/nnano.2015.230

[58] Xiao Z et al. Efficient, high yield perovskite photovoltaic devices grown by interdiffusion of solution-processed precursor stacking layers. Energy & Environmental Science. 2014;**7**:2619

[59] Dou L et al. Solution-processed hybrid perovskite photodetectors with high detectivity. Nature Communications. 2014;**5**:5404

[60] Xu J et al. Perovskite–fullerene hybrid materials suppress hysteresis in planar diodes. Nature Communications. 2015;**6**:7081

[61] Schulz P et al. Charge transfer dynamics between carbon nanotubes and hybrid organic metal halide Perovskite films. Journal of Physical Chemistry Letters. 2016;**7**(3):418-425. DOI: 10.1021/acs.jpclett.5b02721

[62] Ihly R et al. Efficient charge extraction and slow recombination in organic–inorganic perovskites capped with semiconducting single-walled carbon nanotubes. Energy & Environmental Science. 2016;**9**:1439-1449

[63] Hao F, Stoumpos CC, Cao DH, Chang RPH, Kanatzidis MG. Lead-free solid-state organic–inorganic halide perovskite solar cells. Nature Photonics. 2014;**8**:489-494

[64] Kumar MH et al. Lead-free halide Perovskite solar cells with high photocurrents realized through vacancy modulation. Advanced Materials. 2014;**26**:7122-7127

[65] Noel NK et al. Lead-free organic-inorganic tin halide Perovskites for photovoltaic applications. Energy & Environmental Science. 2014;**7**:3061-3068

[66] Ogomi Y et al. $CH_3NH_3Sn_xPb_{(1-x)}I_3$ Perovskite solar cells covering up to 1060 nm. Journal of Physical Chemistry Letters. 2014;**5**:1004-1011

[67] Feng J, Xiao B. Effective masses and electronic and optical properties of nontoxic $MASnX_3$ (X = Cl, Br, and I) Perovskite structures as solar cell absorber: A theoretical study using HSE06. Journal of Physical Chemistry C. 2014;**118**:19655-19660

[68] Babayigit A et al. Assessing the toxicity of Pb- and Sn-based perovskite solar cells in model organism Danio rerio. Nature Scientific Reports. 2016;**6**:18721

[69] Volonakis G et al. Lead-free halide double Perovskites via Heterovalent substitution of Noble metals. Journal of Physical Chemistry Letters. 2016;**7**:1254-1259

[70] Chen Q et al. Under the spotlight: The organic-inorganic hybrid halide perovskite for optoelectronic applications. Nano Today. 2015;**10**:355-396

[71] Tang LC, Chang YC, Huang JY, Lee MH, Chang CS. First principles calculations of linear and second-order optical responses in rhombohedrally distorted perovskite ternary halides, $CsGeX_3$ (X = Cl, Br, and I). Japanese Journal of Applied Physics. 2009;**48**:112402:1-112402:9

[72] Tang LC, Huang JY, Chang CS, Lee MH, Liu LQ. New infrared nonlinear

optical crystal CsGeBr3: Synthesis, structure and powder second-harmonic generation properties. Journal of Physics. Condensed Matter. 2005;**17**:7275-7286

[73] Krishnamoorthy T et al. Lead-free germanium iodide perovskite materials for photovoltaic applications. Journal of Materials Chemistry A. 2015;**3**:23829-23832

[74] Ganose AM, Savory CN, Scanlon DO. Beyond Methylammonium Lead iodide: Prospects for the emergent field of ns 2 containing solar absorbers. Chemical Communications. 2016;**53**(1):20-44. DOI: 10.1039/C6CC06475B

[75] Lyu M et al. Organic–inorganic bismuth (III)-based material: A lead-free, air-stable and solution-processable light-absorber beyond organolead perovskites. Nano Research. 2016;**9**:692-702

[76] Li C et al. Formability of ABX3 (X = F, Cl, Br, I) halide perovskites. Acta Crystallographica. 2008;**64**:702-707

[77] Brandt RE, Stevanović V, Ginley DS, Buonassisi T. Identifying defect-tolerant semiconductors with high minority-carrier lifetimes: Beyond hybrid lead halide perovskites. MRS Communications. 2015;**5**:1-11

[78] Mitzi DB, Brock P. Structure and optical properties of several organic–inorganic hybrids containing corner-sharing chains of bismuth iodide Octahedra. Inorganic Chemistry. 2001;**40**:2096-2104

[79] Kawai T, Ishii A, Kitamura T, Shimanuki S, Iwata M, Ishibashi Y. Optical absorption in band-edge region of $(CH_3NH_3)_3Bi_2I_9$ single crystals. Journal of the Physical Society of Japan. 1996;**65**:1464-1468

[80] Kawai T, Shimanuki S. Optical studies of $(CH_3NH_3)_3Bi_2I_9$ single crystals. Physica Status Solidi B: Basic Solid State Physics. 1993;**177**:K43-K45

[81] Jakubas R, Zaleski J, Sobczyk L. Phase transitions in $(CH_3NH_3)_3$ Bi_2I_9 (MAIB). Ferroelectrics. 1990;**108**:109-114

[82] Eckhardt K et al. Crystallographic insights into $(CH_3NH_3)_3$ (Bi_2I_9): A new lead-free hybrid organic–inorganic material as a potential absorber for photovoltaics. Chemical Communications. 2016;**52**:3058-3060

[83] Park B-W et al. Bismuth based hybrid Perovskites $A_3Bi_2I_9$ (a: Methylammonium or Cesium) for solar cell application. Advanced Materials. 2015;**27**:6806-6813

[84] Hoye RLZ et al. Methylammonium bismuth iodide as a Lead-free, stable hybrid organic-inorganic solar absorber. Chemistry: A European Journal. 2016;**22**:2605-2610

[85] Hamdeh UH et al. Solution-processed BiI3 thin films for photovoltaic applications: Improved carrier collection via solvent annealing. Chemistry of Materials. 2016;**28**:6567-6574

[86] Benabdallah I et al. Lead-free perovskite based bismuth for solar cells absorbers. Journal of Alloys and Compounds. 2019;**773**:796-801

[87] Sanders S et al. Chemical vapor deposition of organic-inorganic bismuth-based Perovskite films for solar cell application. Scientific Reports. 2019;**9**:9774

[88] Ruddlesden SN, Popper P. New compounds of the K2NIF4 type. Acta Crystallographica. 1957;**10**:538-539

[89] Spanopoulos I et al. Uniaxial expansion of the 2D Ruddlesden-Popper Perovskite family for improved environmental stability. Journal of

the American Chemical Society. 2019;**141**(13):5518-5534. DOI: 10.1021/jacs.9b01327

[90] Boyd CC, Cheacharoen R, Leijtens T, McGehee MD. Understanding degradation mechanisms and improving stability of perovskite photovoltaics. Chemical Reviews. 2018;**119**:3418-3451

[91] Gao Y et al. Two-dimensional halide perovskite nanomaterials and heterostructures. Chemical Society Reviews. 2018;**47**:6046-6072

[92] Lan C, Zhou Z, Wei R, Ho JC. Two-dimensional perovskite materials: From synthesis to energy-related applications. Materials Today Energy. 2019;**11**:61-82

[93] Fu Y et al. Broad wavelength Tunable robust lasing from single-crystal nanowires of Cesium Lead halide Perovskites (CsPbX3, X = Cl, Br, I). ACS Nano. 2016;**10**:7963-7972

[94] Zhang Q et al. High-quality whispering- gallery-mode lasing from cesium lead halide perovskite nanoplatelets. Advanced Functional Materials. 2016;**26**:6238-6245

[95] Grancini G, Nazeeruddin MK. Dimensional tailoring of hybrid perovskites for photovoltaics. Nature Reviews Materials. 2019;**4**:4-22

[96] De Bastiani M et al. Ion migration and the role of preconditioning cycles in the stabilization of the J–V characteristics of inverted hybrid perovskite solar cells. Advanced Energy Materials. 2016;**6**:1501453

[97] Sanghadasa M et al. Ultrahigh durability Perovskite solar cells. Nano Letters. 2019;**19**:1251-1259

[98] Asghar MI, Zhang J, Wang H, Lund PD. Device stability of perovskite solar cells—A review. Renewable and Sustainable Energy Reviews. 2017;**77**:131-146

[99] Tsai H et al. High-efficiency two-dimensional ruddlesden-popper perovskite solar cells. Nature. 2016;**536**:312-316. DOI: 10.1038/nature18306

Chapter 3

Synthesis Techniques and Applications of Perovskite Materials

Dinesh Kumar, Ram Sagar Yadav, Monika, Akhilesh Kumar Singh and Shyam Bahadur Rai

Abstract

Perovskite material is a material with chemical formula ABX_3-type, which exhibits a similar crystal structure of $CaTiO_3$. In this material, A and B are metal cations with ionic valences combined to +6, e.g., Li^+:Nb^{5+}; Ba^{2+}:Ti^{4+}; Sr^{2+}:Mn^{4+}; La^{3+}:Fe^{3+}, and X is an electronegative anion with ionic valence (−2), such as O^{2-}, S^{2-}, etc. The properties of a perovskite material strongly depend on the synthesis route of materials. The perovskite materials may be oxides (ABO_3:$CaMnO_3$), halides (ABX_3: X = Cl, Br, I), nitrides (ABN_3:$CaMoN_3$), sulfides (ABS_3:$LaYS_3$), etc., and they may exist in different forms, such as powders, thin films, etc. There are various routes for the synthesis of several perovskites, such as solid-state synthesis, liquid-state synthesis, gas-state synthesis, etc. In this chapter, we discuss various techniques for the synthesis of oxide perovskites in powder form using solid-, liquid-, and gas-state synthesis methods, and we also present an overview on the other type of perovskite materials. The X-ray diffraction, scanning electron microscopy, and optical techniques are used to study the purity of crystallographic phase, morphology, and photoluminescence properties of the perovskite materials. Some applications of different perovskite materials are also discussed.

Keywords: XRD, perovskite, Rietveld refinement, FullProf Suite, lanthanide phosphor

1. Introduction

The general chemical formula of a perovskite material is ABX_3, which contains a crystal structure similar to $CaTiO_3$. It was initially discovered by German geologist Gustav Rose in 1839 in Ural Mountains, and named after Russian mineralogist Lev Perovski [1, 2]. In ABX_3 perovskite, A and B are termed as metal cations having ionic valences combined to +6, e.g., (Li^+:Nb^{5+}; Ba^{2+}:Ti^{4+}; Sr^{2+}:Mn^{4+}; La^{3+}:Fe^{3+}) and X is an electronegative anion with ionic valence −2 such as O^{2-}, S^{2-} etc. [3–6]. The perovskite materials may be oxides, halides, nitrides, sulfides, etc., and they may exist in different forms, such as powders, thin films, etc. [7–10]. The perovskite material has attracted our attention as it can house up a variety of cations at A- and B-sites individually and/or simultaneously along with anions at X-site [11, 12]. The perovskite materials can be classified in ideal and distorted perovskite materials.

IntechOpen

An ideal perovskite material crystallizes into a simple cubic structure with $Pm\bar{3}m$ space group. In the $Pm\bar{3}m$ space group with perovskite structure, A atoms occupy *1(a)* site at (0, 0, 0) and B atoms occupy *1(b)* site at (1/2, 1/2, 1/2) whereas X atoms occupy *3(c)* site at (1/2, 1/2, 0). However, equivalently A, B, and X atoms can also occupy 1(*a*) site at (1/2, 1/2, 1/2), 1(*b*) site at (0, 0, 0) and *3(c)* site at (0, 0, 1/2), respectively, as shown in **Figure 1**. In this figure, A, B, and X are presented in terms of ionic radii [13, 14]. In the unit cell of a perovskite, the cation "B" forms octahedral arrangement with X-anions, i.e., BX_6 and the cation "A" occupies cuboctahedral site with X-anions, i.e., AX_{12}.

The family of perovskite material includes numerous types of oxide forms, such as transition metal oxides with the general formula of ABO_3. The oxide perovskite materials are widely synthesized and are studied for wide applications in various technological fields. In light of these properties, we describe oxide perovskites in more detail.

Victor Moritz Goldschmidt presented an empirical relationship among the ionic radii of A, B, and O, known as tolerance factor (t) to estimate the stability of a perovskite structure. This relation is valid for the relevant ionic radii at room temperature [15]. The numerical value of the tolerance factor can be found by Eq. (1):

$$t = \frac{r_A + r_o}{\sqrt{2}\left(r_B + r_o\right)} \tag{1}$$

where, the term r_A is the ionic radius of cation A and that of r_B is ionic radius of B cation whereas r_O is the ionic radius of oxygen anion (O^{2-}). The ionic radius of A cation is always larger than that of the B cation. The tolerance factor provides an idea about the selection of combination of A and B cations in order to prepare an ideal perovskite material. Eq. (1) can also be expressed in other form, which may be valid for any temperature as given by Eq. (2):

$$t = \frac{d_{A\text{-}o}}{\sqrt{2}\, d_{B\text{-}o}} \tag{2}$$

where d_A-O and d_B-O are average bond-lengths between A-O and B-O, respectively [16].

The distorted perovskite materials are those materials, which crystallize into other than the cubic structures. As far as we know that the perovskite material can accommodate different ions at the A- and B-sites. The variation in the A- and/or B-sites cations causes a variation in the tolerance factor. The variation in tolerance factor leads to a change in the perovskite structure from cubic to non-cubic distorted perovskite structure. For a stable perovskite, the value of tolerance factor should lie in the range of 0.88–1.09 [17]. An ideal perovskite crystal exhibits tolerance factor equal to unity (i.e., $t = 1$). For $t < 1$, the perovskite materials show the rhombohedral or monoclinic structure while in the case of $t > 1$; it reveals tetragonal or orthorhombic structure [18]. Due to distortion in the perovskite system, the BO_6 octahedral led tilted from an ideal situation and causes a change/enhancement in unit cell volume. Thus, the tolerance factor is a measure of the extent of distortion in the perovskite structure. **Figure 2** shows unit cells for some distorted perovskite structures.

There are two general requirements for the formation of a perovskite material, which are given as:

1. **Ionic radii:** the average ionic radii of A- and B-sites cations should be greater than 0.90 Å and 0.51 Å, respectively, and the value of tolerance factor should lie in the range of 0.88–1.09 [19, 20].

DOI: http://dx.doi.org/10.5772/intechopen.86794

2. **Electro-neutrality:** the chemical formula of perovskite material should have neutral balanced charge; consequently, the sum of total charges at A- and B-sites cations must be equal to total charges at O-site (oxygen) of anion(s). A suitable charge distribution is to be achieved in the forms of $A^{3+}B^{3+}O_3$ or $A^{4+}B^{2+}O_3$ or $A^{1+}B^{5+}O_3$ [14, 19].

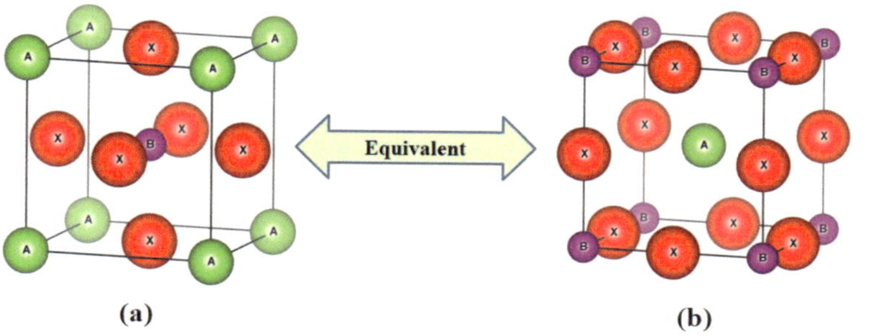

Figure 1.
Molecular structures for ABX_3 perovskite with $Pm\bar{3}m$ space group along with the positions of different atoms in a single unit cell. The figures (a) and (b) are equivalent structures to each other. In the figure (a), A and B take the positions at the corner and body center of the cubic cell, respectively, and X is at the center of face of the unit cell. However, in the figure (b), A and B occupy at body center and the corner of the cubic cell, respectively, and X lies at the center of edge of the unit cell.

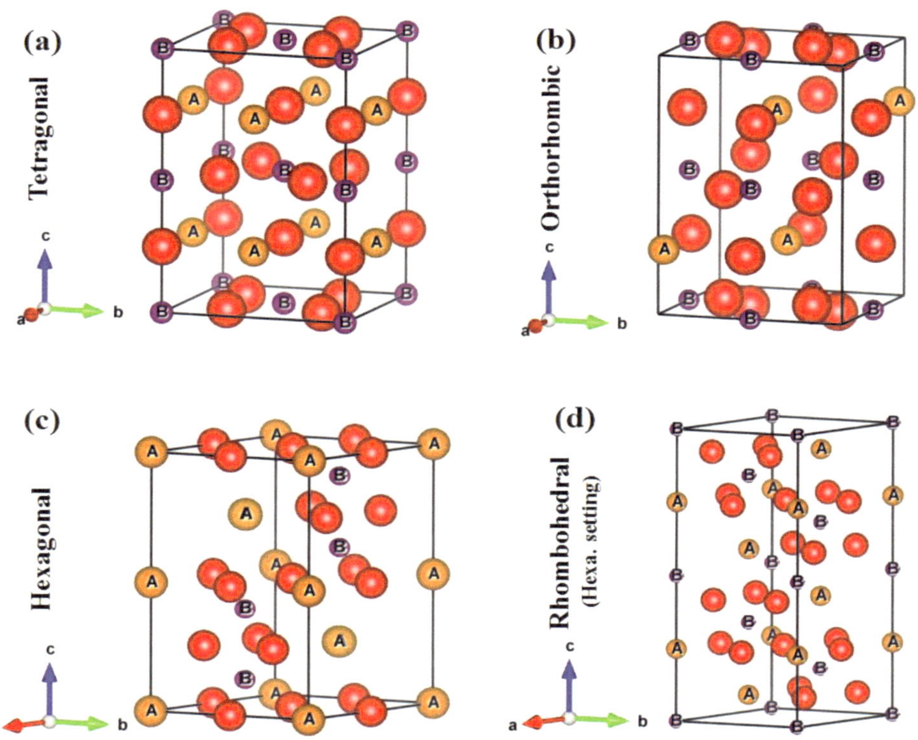

Figure 2.
Different distorted perovskite unit cells; (a) tetragonal, (b) orthorhombic, (c) hexagonal, and (d) rhombohedral. Red spheres stand for oxygen anions.

The pure perovskite materials (ABO_3) do not always provide the desired properties. In order to make them useful, the doping at A- and/or B-sites is required. Doping at A- and/or B-sites improves the properties and also generates very interesting phenomena due to change in crystal structure, bond-lengths, ionic states, etc. The general chemical formula of A- and B-sites after doping in the perovskite matrix may be found in the form of A_1-xA'_xBO_3 ($0 < x < 1$) and AB_1-yB'_yO_3 ($0 < y < 1$), respectively. However, the simultaneous A- and B-sites doped oxide perovskites have general formula A_1-xA'_xB_1-yB'_yO_3. Recently, there are several reports on lanthanide-based rare-earth doped perovskites [21–24]. Initially, we discuss the synthesis process used for the preparation of rare-earth doped perovskite materials alongwith their phase identification, structural analysis by Rietveld refinement of X-ray diffraction (XRD) patterns, morphological and optical properties.

2. Synthesis techniques for perovskite materials

As we know that the physical, chemical, and optical properties of the perovskite materials are strongly synthesis route dependent. One has to choose a suitable synthesis method to obtain the desired properties from the prepared materials. Synthesis techniques also affect crystal structure and morphology of the samples [25]. The synthesis techniques can be divided into three main classes as given below:

1. Solid-state synthesis

2. Liquid-state synthesis

3. Gas-state synthesis

These techniques have their own advantages. Solid-state methods are used to synthesize bulk materials, while liquid-state techniques are used to produce nano-materials. However, gas-state methods are mostly used to fabricate thin films. In this, we will discuss these techniques one by one by taking some suitable examples.

2.1 Solid-state synthesis technique

Solid-state synthesis technique is used to produce polycrystalline materials. It is also known as ceramic method because most of the ceramics are synthesized by this method. This is most widely used technique by researchers. This method requires raw materials in carbonates and/or oxides forms. In this method, the raw materials do not react chemically to each-other at room temperature. When the nixture of raw materials is heated at very high temperatures (i.e., 700–1500°C), the chemical reaction takes place at a significant rate.

2.1.1 Mechanical ball-milling method

It is one of the solid-state synthesis methods used for the production of bulk perovskite materials. Synthesis of perovskites using this method involves grinding, hand mixing, ball milling, and firing of starting materials in many times. In this case, the raw materials used in oxides and/or carbonates forms are grinded, hand mixed, ball milled, and calcined at a high temperature.

Let us describe the whole procedure of this synthesis technique by taking an example of $Ca_{0.97}$-xTiO_3:3Yb^{3+},xBi^{3+} perovskite phosphor. For the synthesis of

$CaCO_3$, TiO_2, Yb_2O_3, and Bi_2O_3 were used as the raw materials. First of all, these materials were weighed in the stoichiometric amounts and grinded and mixed by hand in mortor by using pestle for 1–2 h. After hand mixing and grinding of all the starting materials, the mixture was ball milled in a planetary ball milling system to further get homogeneous mixture in the presence of acetone/alcohol as a mixing media for 4–12 h at a nominal rpm (round per minute) of 25–100 in clockwise and anticlockwise directions. After the ball milling, the homogeneous mixture was dried at ordinary temperature and divided into various parts for calcination at different temperatures from 600 to 1500°C for 4–30 h to optimize the pure phase. These phosphors were then structurally and optically characterized [26]. The perovskites prepared using this method have particles size in the submicrometer range [27]. Some other rare earths doped $R_{0.5}Ba_{0.5}CoO_3$ (R = La, Pr, Nd, Sm, Eu, Gd, Tb, Dy) perovskites were also synthesized using this method [28].

2.1.2 High energy ball-milling method

This technique is very similar to that of mechanical ball-milling technique. Only difference is that this method used very high rpm from few hundreds to few thousands for milling with very small sized balls. This technique uses low temperature for the synthesis of the oxide materials. This method produces generally the nanoparticles. This method only takes metal oxides because there is very high possibility of chemical reactions during high energy ball-milling, which may yield different toxic gases.

Now, let us describe the whole process by considering an example of $0.7BiFeO_3$–$0.3PbTiO_3$ (BF-PT) solid solution [29]. The stoichiometric amounts of Bi_2O_3, Fe_2O_3, PbO, and TiO_2 were taken as raw materials to prepare BF-PT and were mixed using agate mortar and pestle. Then the raw materials were ball-milled for 12–30 h with zirconia balls using alcohol as mixing media in the ratio of 1:10 with the sample and balls at 300–1000 rpm. Here, acetone or acid is not used for the purpose of mixing of raw materials as it may damage the milling jars and O-rings. The material dried at 90°C was annealed at various temperatures (i.e., 400–900°C) to optimize the pure phase of the sample. The XRD pattern of the samples was recorded to check the purity of the synthesized perovskite materials.

For additional characterizations, such as dc poling and dielectric measurements, the annealed sample was mixed in 2% aqueous solution of PVA (polyvinyl alcohol) as binder and a pellet was made. The pellet of the perovskites was fired at 500°C for 10–12 h to burn out the PVA binder. Then the pellet was sintered at quite higher temperatures than annealing temperature to obtain the highly dense perovskite material. To further check the purity of the samples after sintering, the XRD pattern was recorded [30]. Sometimes, this method is also used to decrease the particles size from a micrometric scale to nanometric scale. **Figure 3** represents flow chart of the preparation procedures for the synthesis of perovskite materials using solid-state synthesis technique.

2.2 Liquid-state synthesis technique

The liquid-state synthesis technique is a method of synthesis of nanomaterials. It is most widely used by researchers and scientists for the production of nanoparticles of the oxide materials. In this method, the raw materials may be in the form of nitrates, acetates or oxalates, which may react to each other at an ordinary temperature. Auto-combustion, sol-gel, co-precipitation, etc. are different liquid-state synthesis techniques used for the preparation of perovskite nanomaterials, which are to be described in more detail [19].

2.2.1 Auto-combustion method

The auto-combustion synthesis method is a low-cost and very facile technique for the production of perovskite nanomaterials. In this technique, the starting materials are used in oxalates and/or acetates and/or nitrates forms, which are easily soluble in de-ionized water. It involves some organic fuel, such as urea, citric acid, and glycine to assist the combustion.

Let us describe the process of synthesis of the perovskite materials using auto-combustion method (see **Figure 4(a)**) with the help of an example

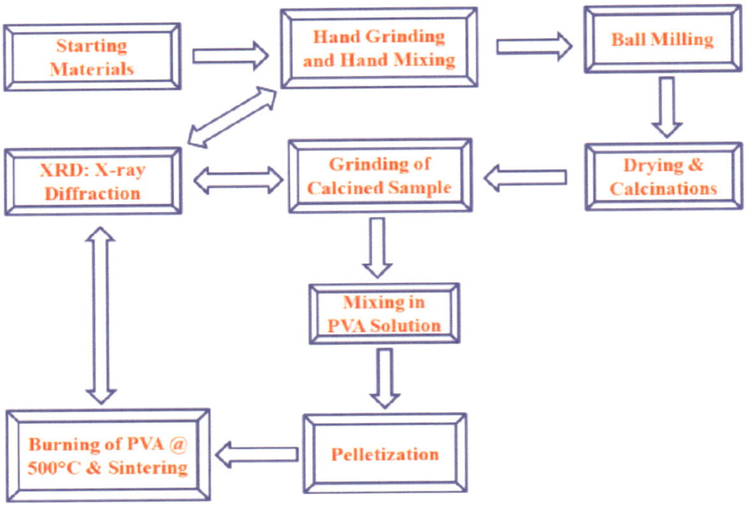

Figure 3.
A representative flow-chart for the synthesis of oxide materials using ball milling technique.

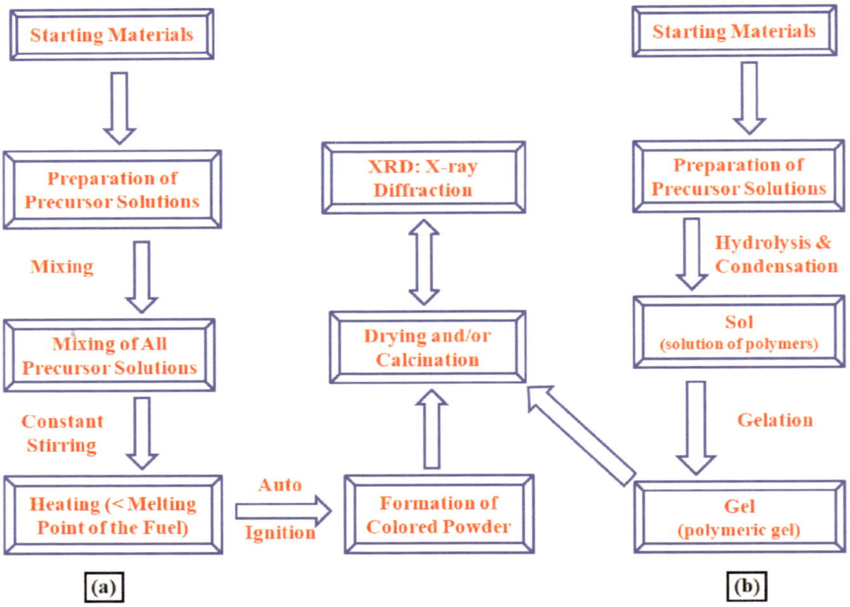

Figure 4.
Schematic flow charts for the synthesis of oxide nanomaterials via (a) auto-combustion and (b) sol-gel methods.

of $La_{0.7}Sr_{0.3}MnO_3$ manganite. Kumar et al. have synthesized $La_{0.7}Sr_{0.3}MnO_3$ perovskite manganite using the starting materials as La_2O_3, $SrCO_3$, and $Mn(CH_3COO)_2.4H_2O$ and glycine was used as fuel. First of all, La_2O_3 and $SrCO_3$ were dissolved in diluted nitric acid to prepare their respective nitrates and then $Mn(CH_3COO)_2.4H_2O$ and glycine were dissolved into distilled water. All the prepared precursor solutions of the materials were dissolved/prepared independently and at the end, they were mixed simultaneously under nonstop stirring in a large beaker and heated on a magnetic stirrer at 175–200°C [31]. After continuous stirring of 6–7 h, the mixed solution became thicker and converted into gel, with further increase in time of stirring, an auto-ignition takes place resulting in flame that evolves huge amount of different gases. During the ignition process, the temperature of the whole mixture may reach upto 800–1000°C for very short time duration. The obtained blackish-brown powder was collected from the beaker and divided into various parts for the calcination at different temperatures. The $La_{0.7}Sr_{0.3}MnO_3$ perovskite comes up with different particles size range. They have also prepared bulk samples of $Ba_{1-x}Sr_xMnO_3$ perovskites using the same method [32]. The $CaTiO_3:Pr^{3+}$, Al^{3+} phosphor was also prepared by Yin et al. using combustion method [33].

2.2.2 Sol-gel method

The sol-gel method is used by most of the chemists for producing nanomaterials. This technique comprises both types of processes (physical and chemical) associated with the following, such as hydrolysis, condensation, polymerization, gelation, drying, and densification [34]. In this technique, the starting materials are used in the form of metal alkoxides. The general chemical formula of metal alkoxides is $M(OR)_x$. Metal alkoxides can be assumed either a derivative of the alcohol ROH, where R is an alkyl group or a derivative of metal hydroxide $M(OH)_x$ [35]. The stoichiometric amounts of metal alkoxides are weighed and dissolved in alcohol or in de-ionized water at a temperature of 60–80°C under steady stirring. It is very important to control the pH value of the metal alkoxides solutions to avoid the formation of the precipitation and to form the homogeneous gel that can be achieved by using basic or acidic solutions. This is known as hydrolysis and condensation leads to form the polymeric chains. The progress of the polymeric chains eventually results to a perceptible improvement in the viscosity of the reaction mixture and the formation of a gel. The obtained gel is to be dried in temperature range between 150 and 200°C to remove the unwanted contents from the gel [35, 36]. After the removal of contents, the obtained gel was annealed at various temperatures in 400–800°C range to obtain the pure phased materials.

Let us discuss the practical processes involve in this technique by taking an example of $La_{0.6}Ca_{0.4}MnO_3$ perovskite. Andrade et al. have synthesized nanotubes and nanoparticles of $La_{0.6}Ca_{0.4}MnO_3$ perovskite manganite using sol-gel method following calcination at different temperatures [37]. They have used stoichiometric amounts of $La(NO_3)_3.6H_2O$, $CaCO_3$ and $Mn(CH_3COO)_2.4H_2O$ for the synthesis of $La_{0.6}Ca_{0.4}MnO_3$ perovskite. The $CaCO_3$ was dissolved in nitric acid to convert into calcium nitrate, while $La(NO_3)_3.6H_2O$ and $Mn(CH_3COO)_2.4H_2O$ were dissolved in distilled water. All the solutions were mixed together in a beaker. The appropriate amount of polyethylene glycol (PEG) was incorporated to the precursor solutions playing the role of polymerizing agent. Then the solution was heated at 70°C for 6 h to complete the polymerization process. At last, the whole solution was converted in yellow viscous gel, which was calcined at different temperatures from 700 to 1000°C. **Figure 4(b)** displays a representative flow chart of the processes involved in the synthesis of perovskite materials using sol-gel method.

2.2.3 Co-precipitation method

The co-precipitation method is also one of the methods used of the production of nanomaterials. This method needs raw materials of metal cations from a general medium and precipitates in the form of oxalates, carbonates, citrates or hydroxides [38–41]. The resultant precipitates are several times washed with distilled water and then obtained product was calcined at various temperatures to acquire the pure phase of the desired materials in the polycrystalline form. This method can yield almost homogeneous polycrystalline powders. The solubility of the used compounds should be very close to each other for a proper precipitation [42]. The precursor solutions are mixed at atomic level resulting lower particles size and it requires very low calcination temperature to get a pure material [43]. The controlling pH of the precursor solution, stirring speed, concentration, and temperature of the mixture are important parameters for the co-precipitation method [44].

The $LaMn_{1-x}Fe_xO_3$ (x = 0, 0.1, 0.2) perovskite synthesized by Geetha et al. is an example of the co-precipitation method [45]. The stoichiometric amounts of $La(NO_3)_3.6H_2O$, $Fe(NO_3)_3.9H_2O$ and $MnCl_2.4H_2O$ were dissolved in distilled water. These solutions were mixed at one platform and stirred continuously at 50°C for 30 min. After this, NaOH solution was added slowly until the pH of solution is attained to 13. The mixed solution of the precursors was again stirred constantly till the formation of black precipitate. The precipitate was collected and washed many times to remove the excess of chlorides and kept in an oven to dry at 50°C. Thus, the final product was calcined at 800°C for 6 h.

It was observed that liquid-state technique produces nanoparticles of perovskite materials at very low temperature. However, the sub-micron sized perovskite materials can be obtained by firing them at higher temperature similar to solid-state technique.

2.3 Gas-state synthesis technique

Gas-state synthesis technique is also used to prepare the nanoparticles. It contains various procedures for the synthesis of powder oxide materials, viz. furnaces, flames, plasmas, and lasers. The basics of thermodynamics and kinetics of the reaction are very similar to each other; however, their reactors are different. These methods provide narrow distribution of the nanoparticles. The dispersion must be reduced for narrow distribution of the nanoparticles as it leads to increase the particles size [25]. Gas-state synthesis technique is a bottom-up method for the synthesis of multifunctional nanoparticles. Method of bottom-up nanofabrication is based on the gathering of nanomaterials from smaller components.

A variety of electronic devices and solar cells of perovskite materials are prepared in terms of thin films via these techniques and they are entirely different from the other synthesis techniques. There are various techniques for the preparation of thin films, such as chemical vapor deposition [46], molecular beam epitaxy [47], laser ablation [48], DC sputtering [49], magnetron sputtering [50], thermal evaporation [51], and electron beam evaporation [52]. The gas-state synthesis method needs particular set up for a good quality of samples to offer the preferred properties. The gas-state synthesis techniques are categorized into three types:

a. Fabrication at the crystallization temperature under a suitable atmosphere condition of temperature.

b. Fabrication in an intermediate temperature range (500–800°C) followed by post-annealing treatment at higher temperatures.

c. Fabrication at a very low substrate temperature followed by post-annealing treatment at very high temperature.

The perovskite materials fabricated using gas-state techniques have variety of applications ranging from optical and anticorrosion coatings to photocatalysts and solar cells, from semiconductor devices to capacitor dielectrics, from bio-implantable devices to chemical reactors and catalysts. One can undoubtedly say that the industrial interest for preparing nanomaterial-based technologies through gas-state synthesis is going to increase in the upcoming years.

3. Structural and optical properties

3.1 Identification of phase purity: X-ray diffraction studies

It is very important to check the phase and its purity of synthesized perovskite materials. Without knowing the phase purity, one cannot come to any conclusion about the properties exhibited by the perovskite materials. The XRD technique is a suitable tool to identify the phase of perovskites. From the XRD data, one can find out relative phase fractions of different phases present in the prepared samples. One can also find out lattice constants, unit cell volume, crystallite size, lattice strain, and theoretical density from Rietveld refinement of the XRD pattern. The XRD technique is also used to optimize the synthesis conditions for the perovskite materials. By matching the XRD pattern of the synthesized material with the standard XRD pattern of the cubic phase of $CaTiO_3$ (CT) perovskite, one can conclude about the phase purity, i.e., whether, the perovskite is pure or has some amount of impurity phase(s) or crystallizes in the distorted perovskite. **Figure 5** shows the XRD pattern of the cubic CT perovskite with JCPDS File No. 75-2100 using X-ray radiation of 1.5406 Å wavelength. The analysis of XRD pattern gives unit cell lattice parameter a = 3.795 Å and unit cell volume V = 54.656 Å^3. An ideal perovskite material displays reflections from all allowed planes of the primitive unit cell. It shows most intense XRD peak for (1 1 0) plane in the 2θ range of 32–34° and first singlet reflection corresponding to (1 0 0) plane between 22 and 24°.

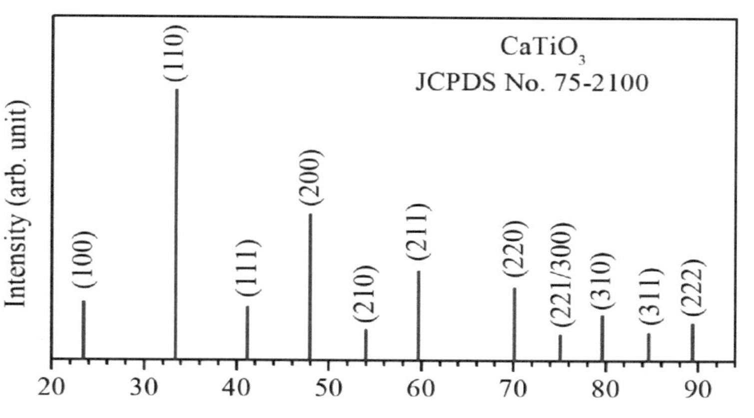

Figure 5.
The standard XRD pattern for the cubic phase of $CaTiO_3$ perovskite.

As we know that the different materials display different XRD patterns as like the finger print of the humans. However, a certain prototype of materials gives the XRD pattern in a well-defined manner. In order to make sure that the synthesized sample is perovskite or not, we have to match the XRD pattern of the synthesized sample with the XRD pattern of CT. If it matches it forms a perovskite phase otherwise not.

Figure 6(a) shows the room-temperature XRD pattern for $BaTiO_3$ (BT) synthesized by mechanical ball-milling method followed by calcination at 950°C for 5 h and it is compared with XRD pattern of CT. The comparison of both the XRD patterns reveals that the XRD pattern of BT matches with that of the CT. This further confirms that BT is a perovskite material. The Bragg's peaks in XRD pattern of BT perovskite is shifted towards lower angle side compared to CT. This indicates that the lattice parameters of BT are larger than that of the CT. Furthermore, the Bragg peak around 22° is asymmetric in lower angle side, which shows doublet nature in the peak, i.e., the lattice plane (1 0 0) of CT split into (0 0 1) and (1 0 0) of BT is shown as inset in **Figure 6(a)**. Similarly, the other Bragg's peaks also show asymmetrical behavior. All these observations reveal that the crystal structure of BT is distorted from an ideal perovskite structure. The Rietveld refinement of XRD pattern of BT reveals that the BT crystallizes into tetragonally distorted structure with *P4mm* space group [53]. It has been also observed that the presence of superlattice reflection(s) in the XRD pattern indicates the formation of distorted perovskite structure from an ideal cubic [54, 55]. The structural analysis of the known or unknown perovskite materials can be identified by Rietveld refinement of the powder XRD pattern with the help of some standard software [56]. Most of researchers use FullProf Suite to identify the structure of various perovskite materials [57]. From Rietveld refinement, one can get lattice parameters, atomic coordinates of the constituent atoms present in the unit cell, average bond-lengths and angles, draw molecular structure to see the surroundings of the A- and B-sites' cations with anions, and many more [29, 33, 58]. One can also find out magnetic structure from the Rietveld refinement of the neutron diffraction data [59].

We performed the Rietveld refinement of XRD pattern of BT using *P4mm* space group of the tetragonal structure. In *P4mm* space group, Ti^{4+}-ions and Ba^{2+}-ions occupy *1(a)* site at (0, 0, 0) and *1(b)* site at (0.5, 0.5, 0.5 + δz), respectively, while, $O^{2-}(1)$ and $O^{2-}(2)$ anions substitute *1(a)* site at (0, 0, 0.5 + δz) and *2(c)* site at (0.5, 0, δz), respectively [53, 60]. The structural analysis provides

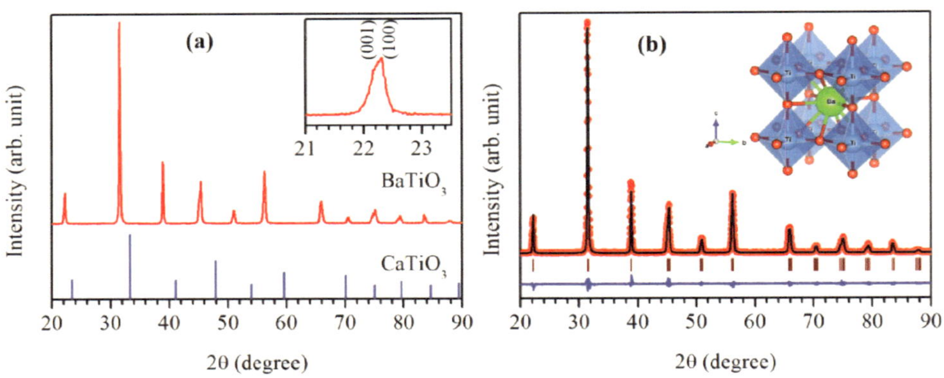

Figure 6.
(a) Room temperature XRD pattern of $BaTiO_3$ compared with the $CaTiO_3$. The inset in (a) shows selected Bragg's peak between 21.0 and 23.5°. (b) Rietveld fit of the XRD pattern of $BaTiO_3$. The inset in (b) shows ball and stick molecular model for the unit cell of $BaTiO_3$ perovskite, where Ba, Ti, and O are present in their atomic sizes.

lattice constants a = b = 4.0005(1) Å and c = 4.0255(2) Å and unit cell volume V = 64.423(4) $Å^3$ with tetragonality (c/a) of 1.006 close to the earlier reported value [60]. **Figure 6(b)** shows Rietveld fit of the XRD pattern for the BT perovskite. In this figure, the scattered dots are used to present experimental XRD pattern and continuous line over experimental pattern is to show the simulated XRD pattern. Lower continuous curve shows difference between experimental and simulated XRD patterns. The vertical bars represent positions of the Bragg's reflection. The inset of **Figure 6(b)** demonstrates ball and stick molecular model for the unit cell of *P4mm* space group for BT in terms of atomic sizes. It is clear that Ba atom forms octahedral with oxygen atoms (BaO_6) and Ti atom forms cuboctahedral with oxygen atoms (TiO_{12}).

3.2 Morphological studies

The scanning electron microscopy (SEM) and transmission electron microscopy (TEM) can be used to study the microstructure nature of the perovskite materials. We have recorded SEM micrographs of $Nd_{0.4}Sr_{0.6}MnO_3$ manganite samples calcined at 800°C and 1200°C synthesized by auto-combustion method using FEI, Nova Nano SEM for the $Nd_{0.4}Sr_{0.6}MnO_3$ perovskite and they are shown in **Figure 7**. The average values of the particle size were analyzed by ImageJ software and they are found to be 50 and 425 nm for the $Nd_{0.4}Sr_{0.6}MnO_3$ perovskites calcined at 800°C and 1200°C, respectively. This clearly shows that the average particle size improves on increasing the calcination temperature. It confirms that the sample calcined at 800°C produces nanomaterial whereas that of at 1200°C gives bulk material.

3.3 Optical studies

Figure 8 shows photoluminescence (PL) excitation and PL emission spectra of the $Nd_{0.4}Sr_{0.6}MnO_3$ manganite calcined at 800°C and 1200°C. **Figure 8(a)** shows the excitation spectra of the $Nd_{0.4}Sr_{0.6}MnO_3$ manganite in both the cases and contains an intense peak at 355 nm [45]. It was monitored at an emission wavelength of 646 nm. The intensity of bulk material is two times larger than the nanomaterial. When both the materials were excited with 355 nm they give strong red color at 646 nm as shown in **Figure 8(b)**. It also contains weak peaks at 483 and 582 nm. The emission intensity obtained in the case of bulk material is further two times larger than the nanomaterial. This is due to the increase in crystallinity and particles size of the materials. Thus, the calcination affects the morphology and optical properties of perovskites even though the perovskite material was synthesized by auto-combustion method.

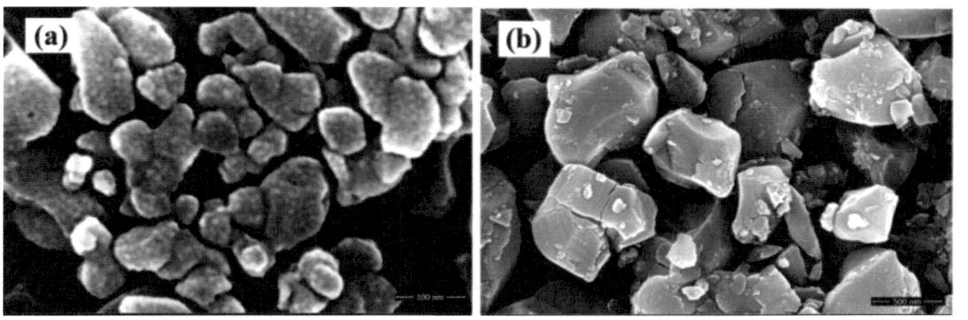

Figure 7.
SEM micrographs of $Nd_{0.4}Sr_{0.6}MnO_3$ manganite calcined at (a) 800°C and (b) 1200°C temperatures.

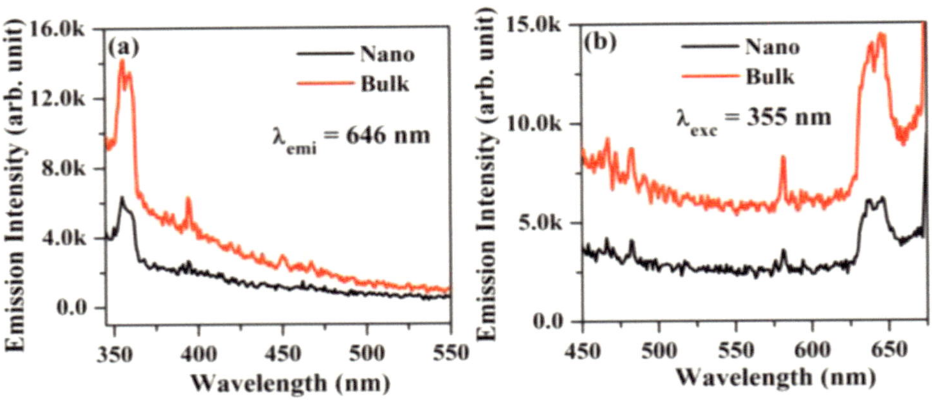

Figure 8.
(a) PL excitation and (b) PL emission spectra of the $Nd_{0.4}Sr_{0.6}MnO_3$ manganite calcined at 800°C and 1200°C temperatures.

4. Applications of perovskite materials

The perovskite materials are extensively studied by researchers due to their attractive properties. The perovskite materials have wide applications in various fields, which are listed below:

1. Photocatalytic activity; e.g., $LaFeO_3$ [61].
2. Photovoltaic solar cells; e.g., $LaVO_3$ [62].
3. Phosphor materials in photoluminescence; e.g., $Ho^{3+}/Yb^{3+}/Mg^{2+}$ doped $CaZrO_3$ [63].
4. Solid oxide fuel cells; e.g., $Gd_{0.7}Ca_{0.3}Co_{1-y}Mn_yO_3$ [64].
5. Sensors and actuators; e.g., $PbZr_xTi_1$-xO_3 [65].
6. Magnetic memory devices; e.g., $Pt/La_2Co_{0.8}Mn_{1.2}O_6/Nb:SrTiO_3$ [66].
7. Magnetic field sensors; e.g., $La_{0.67}Sr_{0.33}MnO_3$ and $La_{0.67}Ba_{0.33}MnO_3$ [67].
8. Electric field effect devices; e.g., hetrostructure of $Pb(Zr_{0.2}Ti_{0.8})O_3/La_{0.8}Ca_{0.2}MnO_3$ [68].
9. Ferroelectric and piezoelectric devices; e.g., $BaTiO_3$, $PbTiO_3$ [69].
10. Semiconducting electronic devices; e.g., $La_{0.7}Ca_{0.3}MnO_3/SrTiO_3/La_{0.7}Ce_{0.3}MnO_3$ [70].
11. High dielectric constant; e.g., Bi_1-xSr_xMnO_3 (x = 0.4, 0.5) [71].
12. High temperature superconductor; e.g., $BaPb_1$-xBi_xO_3 [72].
13. Hypothermia; e.g., $La_{0.7}Sr_{0.3}MnO_3$ [73].
14. Supercapacitor; e.g., $KNi_{0.8}Co_{0.2}F_3$ [74].

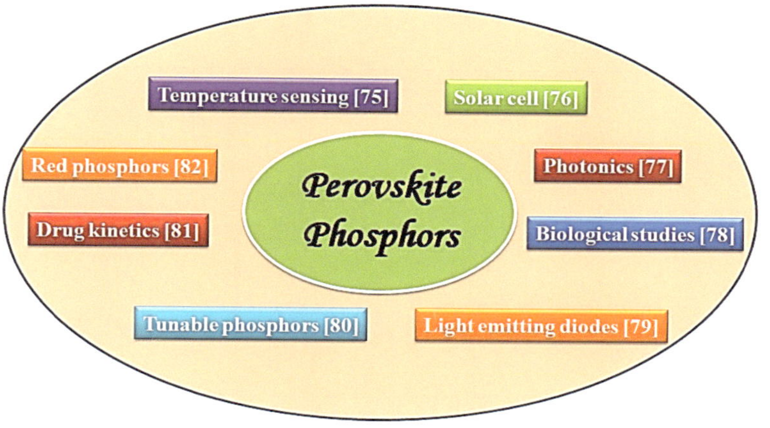

Figure 9.
Block diagram of the lanthanide based perovskite phosphor materials including their applications in various fields [75–82].

The lanthanide based perovskite phosphors are one of them and have unique photolumincence properties. The lanthanide ions are known for their narrow and sharp band emissions. They give emissions in the entire range of visible spectrum along with ultraviolet (UV) and near infrared (NIR) regions. These emissions are observed due to presence of meta-stable energy levels in the lanthanide ions. These levels have long lifetime and are responsible for very strong emissions. The lanthanide ions possess upconversion, downconversion, and quantum cutting phenomena. In upconversion, the two or more than two low energy photons are converted into high energy photons. When a high energy photon is converted into low energy photons it is termed as downconversion process. However, in quantun cutting, a high energy photon is converted into two low energy NIR photons. These processes lead numerous technological applications in different fields, such as red phosphors, light emitting diodes (LEDs), photonics, solar cells, tunable phosphors, temperature sensing, biological studies, drug kinetics, etc. [63, 75–82]. **Figure 9** shows a block diagram for the lanthanide based perovskite phosphors, which shows their applications in various fields.

5. Conclusions

This chapter summarizes the basics of the perovskite structure, its stability and distortion. We have discussed the novelty of the perovskite materials, which accommodate different cations at A- and/or B-sites individually and/or simultaneously. We have also discussed various routes such as solid, liquid and gas-state synthesis for the preparation of perovskite materials mostly in the oxide powder forms. We have also briefly described the phase identification of the perovskites and their structural analysis using Rietveld refinement of the XRD data by taking an example of tetragonal $BaTiO_3$ perovskite. The morphological and optical studies were also incorporated. We have also briefly listed various applications of perovskite materials including lanthanide based perovskite phosphors in various fields.

Conflict of interest

The authors declare no conflict of interest in the chapter.

Author details

Dinesh Kumar[1], Ram Sagar Yadav[2*], Monika[3], Akhilesh Kumar Singh[1] and Shyam Bahadur Rai[3]

1 School of Materials Science and Technology, Indian Institute of Technology, Banaras Hindu University, Varanasi, Uttar Pradesh, India

2 Department of Zoology, Institute of Science, Banaras Hindu University, Varanasi, Uttar Pradesh, India

3 Department of Physics, Institute of Science, Banaras Hindu University, Varanasi, Uttar Pradesh, India

*Address all correspondence to: ramsagaryadav@gmail.com

References

[1] Roth RS. Classification of perovskite and other ABO_3-type compounds. Journal of Research of the National Bureau of Standards. 1957;**58**:75-88

[2] Yashima M, Ali R. Structural phase transition and octahedral tilting in the calcium titanate perovskite $CaTiO_3$. Solid State Ionics. 2009;**180**:120-126

[3] Weis RS, Gaylord TK. Lithium niobate: Summary of physical properties and crystal structure. Applied Physics A: Materials Science & Processing. 1985;**37**:191-203

[4] Kwei GH, Lawson AC, Billinge SJL, Cheong SW. Structures of the ferroelectric phases of barium titanate. The Journal of Physical Chemistry. 1993;**97**:2368-2377

[5] Sakai H, Ishiwata S, Okuyama D, Nakao A, Nakao H, Murakami Y, et al. Electron doping in the cubic perovskite $SrMnO_3$: Isotropic metal versus chainlike ordering of Jahn-Teller polarons. Physical Review B. 2010;**82**:180409R–180412R

[6] Wang Y, Zhu J, Zhang L, Yang X, Lu L, Wang X. Preparation and characterization of perovskite $LaFeO_3$ nanocrystals. Materials Letters. 2006;**60**:1767-1770

[7] Kubicek M, Bork AH, Rupp JLM. Perovskite oxides—A review on a versatile material class for solar-to-fuel conversion processes. Journal of Materials Chemistry A. 2017;**5**:11983-12000

[8] Protesescu L, Yakunin S, Bodnarchuk MI, Krieg F, Caputo R, Hendon CH, et al. Nanocrystals of cesium lead halide perovskites ($CsPbX_3$, X = Cl, Br, and I): Novel optoelectronic materials showing bright emission with wide color gamut. Nano Letters. 2015;**15**:3692-3696

[9] Perez RS, Cerqueira TFT, Korbel S, Botti S, Marques MAL. Prediction of stable nitride perovskites. Chemistry of Materials. 2015;**27**:5957-5963

[10] Kuhar K, Crovetto A, Pandey M, Thygesen KS, Seger B, Vesborg PCK, et al. Sulfide perovskites for solar energy conversion applications: Computational screening and synthesis of the selected compound $LaYS_3$. Energy & Environmental Science. 2017;**10**:2579-2593

[11] Kansara SB, Dhruv D, Kataria B, Thaker CM, Rayaprol S, Prajapat CL, et al. Structural, transport and magnetic properties of monovalent doped $La_{1-x}Na_xMnO_3$ manganites. Ceramics International. 2015;**41**:7162-7173

[12] Pandey R, Pillutla RK, Shankar U, Singh AK. Absence of tetragonal distortion in $(1-x)SrTiO_{3-x}Bi(Zn_{1/2}Ti_{1/2})O_3$ solid solution. Journal of Applied Physics. 2013;**113**:184109-184114

[13] Ali Z, Ahmad I, Amin B, Maqbool M, Murtaza G, Khan I, et al. Theoretical studies of structural and magnetic properties of cubic perovskites $PrCoO_3$ and $NdCoO_3$. Physica B. 2011;**406**:3800-3804

[14] Johnsson M, Lemmens P. Crystallography and chemistry of perovskites. In: Kronmuller H, Parkin S, editors. Handbook of Magnetism and Advanced Magnetic Materials. Vol. 4. USA: John Wiley & Sons; 2005. pp. 1-11

[15] Goldschmidt VM. Die Gesetze der Krystallochemie. Naturwissenschaften. 1926;**14**:477-485

[16] Kumar D, Verma NK, Singh CB, Singh AK. Evolution of structural characteristics of $Nd_{0.7}Ba_{0.3}MnO_3$ perovskite manganite as a function of crystallite size. AIP Conference Proceedings. 2018 2009:020013-4

[17] Duan R. High curie temperature bismuth- and indium-substituted lead titanate [thesis]. Georgia Institute of Technology: School of Materials Science and Engineering; 2004

[18] Kumar D, Singh AK. Investigation of structural and magnetic properties of $Nd_{0.7}Ba_{0.3}Mn_{1-x}Ti_xO_3$ (x = 0.05, 0.15 and 0.25) manganites synthesized through a single-step process. Journal of Magnetism and Magnetic Materials. 2019;**469**:264-273

[19] Atta NF, Galal A, El-Ads EH. Synthesis, characterization and applications. In: Pan L, Zhu G, editors. Perovskite Nanomaterials. London: IntechOpen; 2016. pp. 107-151. Ch. 4

[20] Porta P, De RS, Faticanti M, Minelli G, Pettiti I, Lisi L, et al. Perovskite-type oxides: I. Structural, magnetic, and morphological properties of $LaMn_{1-x}Cu_xO_3$ and $LaCo_{1-x}Cu_xO_3$ solid solutions with large surface area. Journal of Solid State Chemistry. 1999;**146**:291-304

[21] Chen M, Zhang H, Liu T, Jiang H, Chang A. Preparation, structure and electrical properties of $La_{1-x}Ba_xCrO_3$NTC ceramics. Journal of Materials Science: Materials in Electronics. 2017;**28**:18873-18188

[22] Yang J, Ma YQ, Zhang RL, Zhao BC, Ang R, Song WH, et al. Structural, transport, and magnetic properties in the Ti-doped manganites $LaMn_{1-x}Ti_xO_3$ ($0 \leq x \leq 0.2$). Solid State Communications. 2005;**136**:268-272

[23] Kameli P, Salamati H, Heidarian A, Bahrami H. Ferromagnetic insulating and reentrant spin glass behavior in Mg doped $La_{0.75}Sr_{0.25}MnO_3$ manganites. Journal of Non-Crystalline Solids. 2009;**355**:917-921

[24] Choudhary N, Verma MK, Sharma ND, Sharma S, Singh D. Correlation between magnetic and transport properties of rare earth doped perovskite manganites $La_{0.6}R_{0.1}Ca_{0.3}MnO_3$ (R = La, Nd, Sm, Gd, and Dy) synthesized by Pechini process. Materials Chemistry and Physics. 2020;**242**:122482 pp. 11

[25] Ring TA. Fundamentals of Ceramic Powder Processing and Synthesis: Ceramic Powder Synthesis. United States: Academic Press; 1996

[26] Lin L-T, Chen J-Q, Deng C, Tang L, Chen D-J, Meng J-X, et al. Broadband near-infrared quantum-cutting by cooperative energy transfer in Yb^{3+}–Bi^{3+} co-doped $CaTiO_3$ for solar cells. Journal of Alloys and Compounds. 2015;**640**:280-284

[27] Pandey R, Tiwari A, Upadhyay A, Singh AK. Phase coexistence and the structure of the morphotropic phase boundary region in (1-x)$Bi(Mg_{1/2}Zr_{1/2})O_3$-x$PbTiO_3$ piezoceramics. Acta Materialia. 2014;**76**:198-206

[28] Troyanchuk IO, Kasper NV, Khalyavin DD, Szymczak H, Szymczak R, Baran M. Magnetic and electrical transport properties of orthocobaltites $R_{0.5}Ba_{0.5}CoO_3$ (R = La, Pr, Nd, Sm, Eu, Gd, Tb, Dy). Physical Review B. 1998;**58**:2418-2421

[29] Zhang L, Xu Z, Cao L, Yao X. Synthesis of BF-PT perovskite powders by high-energy ball milling. Materials Letters. 2007;**61**:1130-1133

[30] Upadhyay A, Pandey R, Singh AK. Origin of ferroelectric P-E loop in cubic compositions and structure of poled (1-x)$Bi(Mg_{1/2}Zr_{1/2})O_3$-x$PbTiO_3$ piezoceramics. Journal of the American Ceramic Society. 2017;**100**:1743-1750

[31] Kumar D, Verma NK, Singh CB, Singh AK. Crystallite size strain analysis of nanocrystalline $La_{0.7}Sr_{0.3}MnO_3$ perovskite by Williamson-Hall plot method. AIP Conference Proceedings. 2018 1942:050024-4

[32] Kumar D, Singh CB, Verma NK, Singh AK. Synthesis and structural investigations on multiferroic $Ba_{1-x}Sr_xMnO_3$ perovskite manganites. Ferroelectrics. 2017;**518**:191-195

[33] Yin S, Chen D, Tang W. Combustion synthesis and luminescent properties of $CaTiO_3$:Pr, Al persistent phosphors. Journal of Alloys and Compounds. 2007;**441**:327-331

[34] Brinker CJ, Scherer GW. Sol-Gel Science: The Physics and the Chemistry of Sol-Gel Processing. London: Academic Press Inc.; 1990

[35] Rahaman MN. Ceramic Processing and Sintering. 2nd ed. New York: Marcel Dekker Inc.; 2003

[36] Rajaeiyan A, Mohagheghi MMB. Comparison of sol-gel and co-precipitation methods on the structural properties and phase transformation of γ and α-Al_2O_3 nanoparticles. Advanced Manufacturing. 2013;**1**:176-182

[37] Andrade VM, Vivas RJC, Pedro SS, Tedesco JCG, Rossi AL, Coelho AA, et al. Magnetic and magnetocaloric properties of $La_{0.6}Ca_{0.4}MnO_3$ tunable by particle size and dimensionality. Acta Materialia. 2016;**102**:49-55

[38] Pei RR, Chen X, Suo Y, Xiao T, Ge QQ, Yao HC, et al. Synthesis of $La_{0.85}Sr_{0.15}Ga_{0.8}Mg_{0.2}O_{3}$-δ powder by carbonate co-precipitation combining with azeotropic-distillation process. Solid State Ionics. 2012;**219**:34-40

[39] Uskokovic V, Drofenik M. Four novel co-precipitation procedures for the synthesis of lanthanum-strontium manganites. Materials and Design. 2007;**28**:667-672

[40] Cho TH, Shiosaki Y, Noguchi H. Preparation and characterization of layered $LiMn_{1/3}Ni_{1/3}Co_{1/3}O_2$ as a cathode material by an oxalate co-precipitation method. Journal of Power Sources. 2006;**159**:1322-1327

[41] Wei Y, Han B, Hu X, Lin Y, Wang X, Deng X. Synthesis of Fe_3O_4 nanoparticles and their magnetic properties. Procedia Engineering. 2012;**27**:632-637

[42] West AR. Solid State Chemistry and its Applications. USA: John Wiley & Sons; 2005

[43] Gaikwad AB, Navale SC, Samuel V, Murugan AV, Ravi V. A co-precipitation technique to prepare $BiNbO_4$, $MgTiO_3$ and $Mg_4Ta_2O_9$ powders. Materials Research Bulletin. 2006;**41**:347-353

[44] Zawrah MF, Hamaad H, Meky S. Synthesis and characterization of nano $MgAl_2O_4$ spinel by the co-precipitated method. Ceramics International. 2007;**33**:969-978

[45] Geetha N, Senthil KV, Prakash D. Synthesis and characterization of $LaMn_{1-x}Fe_xO_3$ (x = 0, 0.1, 0.2) by coprecipitation route. Journal of Physical Chemistry & Biophysics. 2018;**8**:273-278

[46] Kwak BS, Zhang K, Boyd EP, Erbil A, Wilkens BJ. Metalorganic chemical vapor deposition of $BaTiO_3$ thin films. Journal of Applied Physics. 1991;**69**:767-772

[47] Yu Z, Ramdani J, Curless JA, Finder JM, Overgaard CD, Droopad R, et al. Epitaxial perovskite thin films grown on silicon by molecular beam epitaxy. Journal of Vacuum Science and Technology B. 2000;**18**:1653-1657

[48] Imai T, Okuyama M, Hamakawa Y. $PbTiO_3$ thin films deposited by laser ablation. Japanese Journal of Applied Physics. 1991;**30**:2163-2166

[49] Bangchao Y, Wang JY, Jia YM, Yongjie H. Preparation of $PbTiO_3$ thin film by dc single-target magnetron

sputtering. Proceedings of the SPIE. 1991;**1519**:725-728

[50] Lu CJ, Shen HM, Wang YN. Preparation and crystallization of $Pb(Zr_{0.95}Ti_{0.05})O_3$ thin films deposited by radio-frequency magnetron sputtering with a stoichiometric ceramic target. Applied Physics A: Materials Science and Processing. 1998;**67**:253-258

[51] Li Y, Xu X, Wang C, Wang C, Xie F, Yang J, et al. Investigation on thermal evaporated $CH_3NH_3PbI_3$ thin films. AIP Advances. 2015;**5**:097111-097116

[52] Pae SR, Byun S, Kim J, Kim M, Gereige I, Shin B. Improving uniformity and reproducibility of hybrid perovskite solar cells via a low-temperature vacuum deposition process for NiO_x hole transport layers. ACS Applied Materials & Interfaces. 2018;**10**:534-540

[53] Xiao CJ, Jin CQ, Wang XH. Crystal structure of dense nanocrystalline $BaTiO_3$ ceramics. Materials Chemistry and Physics. 2008;**111**:209-212

[54] Geller S. Crystal structure of gadolinium orthoferrite $GdFeO_3$. The Journal of Chemical Physics. 1956;**24**:1236-1239

[55] Jakymiw C, Vocadlo L, Dobson DP, Bailey E, Thomson AR, Brodholt JP, et al. The phase diagrams of $KCaF_3$ and $NaMgF_3$ by *ab initio* simulations. Physics and Chemistry of Minerals. 2018;**45**:311-322

[56] Young RA. The Rietveld Method. International Union of Crystallography: Oxford University Press; 1993

[57] Carvajal JR. 'FULLPROF' Program: Rietveld Pattern Matching Analysis of Powder Patterns-ILL. Grenoble; 1990

[58] Yadav RS, Kumar D, Singh AK, Rai E, Rai SB. Effect of Bi^{3+} ion on upconversion-based induced optical heating and temperature sensing characteristics in the Er^{3+}/Yb^{3+} co-doped La_2O_3 nano-phosphor. RSC Advances. 2018;**8**:34699-34711

[59] Carvajal JR. Recent advances in magnetic structure determination by neutron powder diffraction. Physica B. 1993;**192**:55-69

[60] Villars P. Pauling file in: Inorganic Solid Phases. Springer Materials (online database). Springer Materials. $BaTiO_3$ tetragonal ($BaTiO_3$ rt) Crystal Structure. Heidelberg: Springer. Available from: https://materials.springer.com/isp/crystallographic/docs/sd_1626689

[61] Afifah N, Saleh R. Enhancement of photocatalytic activities of perovskite $LaFeO_3$ composite by incorporating nanographene platelets. IOP Conference Series: Materials Science and Engineering. 2017;**188**:012054-012058

[62] Wang L, Li Y, Bera A, Ma C, Jin F, Yuan K, et al. Device performance of the Mott insulator $LaVO_3$ as a photovoltaic material. Physical Review Applied. 2015;**3**:064015-064029

[63] Maurya A, Yadav RS, Yadav RV, Rai SB, Bahadur A. Enhanced green upconversion photoluminescence from Ho^{3+}/Yb^{3+} co-doped $CaZrO_3$ phosphor via Mg^{2+} doping. RSC Advances. 2016;**6**:113469-113477

[64] Skinner SJ. Recent advances in perovskite-type materials for solid oxide fuel cell cathodes. International Journal of Inorganic Materials. 2001;**3**:113-121

[65] Uchino K. Glory of piezoelectric perovskites. Science and Technology of Advanced Materials. 2015;**16**:046001-0460016

[66] Mir LL, Frontera C, Aramberri H, Bouzehouane K, Fernandez JC, Bozzo B, et al. Anisotropic sensor and memory device with a ferromagnetic tunnel

barrier as the only magnetic element. Scientific Reports. 2018;**8**:861-870

[67] Xu Y, Memmert U, Hartmann U. Magnetic field sensors from polycrystalline manganites. Sensors and Actuators A. 2001;**91**:26-29

[68] Zhao T, Ogale SB, Shinde SR, Ramesh R, Droopad R, Yu J, et al. Colossal magnetoresistive manganite-based ferroelectric field-effect transistor on Si. Applied Physics Letters. 2004;**84**:750-752

[69] Nuraje N, Su K. Perovskite ferroelectric nanomaterials. Nanoscale. 2013;**5**:8752-8780

[70] Mitra C, Raychaudhuri P, Kobernik G, Dorr K, Muller KH, Schultz L, et al. p-n diode with hole- and electron-doped lanthanum manganites. Applied Physics Letters. 2001;**79**:2408-2410

[71] Munoz JLG, Frontera C, Murias BR, Mira J. Dielectric properties of $Bi_{1-x}Sr_xMnO_3$ (x = 0.40, 0.50) manganites: Influence of room temperature charge order. Journal of Applied Physics. 2009;**105**:084116-084120

[72] Sleight AW, Gillson JL, Bierstedt PE. High-temperature superconductivity in the $BaPb_{1-x}Bi_xO_3$ system. Solid State Communications. 1993;**88**:841-842

[73] Manh DH, Phong PT, Nam PH, Tung DK, Phuc NX, Lee IJ. Structural and magnetic study of $La_{0.7}Sr_{0.3}MnO_3$ nanoparticles and AC magnetic heating characteristics for hyperthermia applications. Physica B. 2014;**444**:94-102

[74] Ding R, Li X, Shi W, Xu Q, Han X, Zhou Y, et al. Perovskite $KNi_{0.8}Co_{0.2}F_3$ nanocrystals for supercapacitors. Materials Chemistry A. 2017;**5**:17822-17827

[75] Mahata MK, Koppe T, Mondal T, Brusewitz C, Kumar K, Rai VK, et al. Incorporation of Zn^{2+} ions into $BaTiO_3$:Er^{3+}/Yb^{3+} nanophosphor: An effective way to enhance upconversion, defect luminescence and temperature sensing. Physical Chemistry Chemical Physics. 2015;**17**:20741-20753

[76] Ho WJ, Li GY, Liu JJ, Lin ZX, You BJ, Ho CH. Photovoltaic performance of textured silicon solar cells with $MAPbBr_3$ perovskite nanophosphors to induce luminescent down-shifting. Applied Surface Science. 2018;**436**:927-933

[77] Singh DK, Manam J. Structural and photoluminescence studies of red emitting $CaTiO_3$:Eu^{3+} perovskite nanophosphors for lighting applications. Journal of Materials Science: Materials in Electronics. 2016;**27**:10371-10381

[78] Kumar KN, Vijayalakshmi L, Choi J. Investigation of upconversion photoluminescence of Yb^{3+}/Er^{3+}:$NaLaMgWO_6$ noncytotoxic double-perovskite nanophosphors. Inorganic Chemistry. 2019;**58**:2001-2011

[79] Tian Y, Zhou C, Worku M, Wang X, Ling Y, Gao H, et al. Highly efficient spectrally stable red perovskite light-emitting diodes. Advanced Materials. 2018;**30**:1707093-1707097

[80] Jain N, Singh RK, Sinha S, Singh RA, Singh J. Color tunable emission through energy transfer from Yb^{3+} co-doped $SrSnO_3$:Ho^{3+} perovskite nano-phosphor. Applied Nanoscience. 2018;**8**:1267-1278

[81] Li X, Zhang Q, Ahmad Z, Huang J, Ren Z, Weng W, et al. Near-infrared luminescent $CaTiO_3$:Nd^{3+} nanofibers with tunable and trackable drug release kinetics. Journal of Materials Chemistry B. 2015;**3**:7449-7456

[82] Itoh S, Toki H, Tamura K, Kataoka F. A new red-emitting phosphor, $SrTiO_3$:Pr^{3+} for low-voltage electron excitation. Japanese Journal of Applied Physics. 1999;**38**:6387-6391

Section 2

Applications of Perovskite in Pollutant Degradation and Energy Storage

Chapter 4

Developing Functionality in Perovskites from Abatement of Pollutants to Sustainable Energy Conversion and Storage

Andrea Bedon and Antonella Glisenti

Abstract

Different functionalities can be developed and optimised in perovskites by means of different approaches: doping, nanocomposition and exsolution. In our research, these strategies are used with the aim of a sustainable development. Our objectives are in developing perovskite-based materials for catalysis and electro-catalysis. In particular, we successfully obtained catalysts for abatement of pollutants in three-way catalysts, catalysts for dry reforming of methane and electro-catalysts for solid oxide cells (solid oxide fuel cells and solid oxide electrolysers). In all cases, the corresponding device and experimental set-up was optimised. The catalysts and electro-catalysts we considered are free of noble metals and with minimum amount of critical raw materials. The preparation procedures are by wet chemistry, highly reproducible and up-scalable.

Keywords: catalysis, electrocatalysis, nanocomposition, doping, oxygen exchange, solid oxide cells, three-way catalysts, sustainable development, energy storage and conversion, pollutants abatement

1. Introduction

Several materials and devices are very efficient in terms of catalytic and electro-catalytic performance because of the presence of noble metals. Due to the cost and difficulty of supply, the European Community decided that the minimization of noble metals use is a fundamental step for economic and social development. Perovskites are good candidates for substituting noble metals in devices because of their versatility and low economic and environmental impact. The perovskites we focus on are oxides of the type ABO_3 (with A = alkali or alkali earth or rare earth and B = transition metal, usually) [1]. These are very stable oxides, characterised by great versatility in composition, tolerance to structural deformation and, typically, low cost. The functionality of perovskites can be enhanced using different approaches. In this chapter, several strategies for the enhancement of catalytic and electro-catalytic performance are described and compared. Numerous devices require materials characterised by similar properties such as the capability to dissociatively coordinate molecules, the ionic mobility and exchange capability, the thermal and morphological stability, etc. Different methodologies allow to build functionality in perovskite-type oxides; among these, doping, nanocomposition and exosolution are the more relevant, being more easy to control and characterised by higher effect. The doping in A-site is fundamental to affect the oxygen mobility and ion conductivity but can also enforce a redox couple. The deformation caused by this doping may have relevant effects on electronic conductivity and mobility. Doping in B-site is a precious help in creating redox behaviour and, consequently, catalytic activity and electronic conductivity. When different properties are required, it can be easier to model the behaviour through nanocomposition. The synergy between supported and supporting phases, in fact, allows obtaining higher performances when compared to the sum of the two. At this purpose, several aspects are relevant in affecting the behaviour of the interfaces and they will be considered. The rearrangement of perovskite surface is very relevant in determining the interaction between molecules (reactants, intermediates and products), and exosolution can be very effective in inducing the reconstruction of active surfaces.

These procedures for enhancing and tuning the catalytic and electro-catalytic performances of perovskite-based material will be compared and evaluated taking into consideration three applications and devices: solid oxide cells (fuel cells—SOFCs—and electrolyzers—SOECs), catalysis for abatement of pollutants (three-way catalysts, TWC) and for green hydrogen/green fuel generation (dry reforming of methane, DRM).

2. Developing functionality in perovskites for a sustainable development: experimental

2.1 Synthesis

The samples are always prepared by citrate method [2] starting from a solution of precursors to which citric acid is added. The molar ratio of citric acid

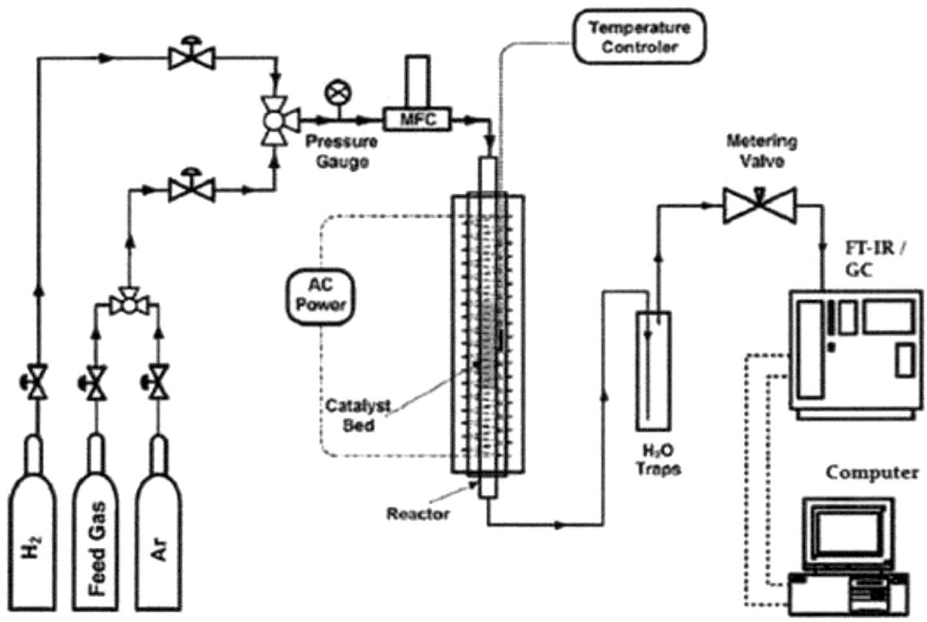

Figure 1.
Experimental set-up used for the catalytic activity measurements.

monohydrate was 1.9:1 with respect to the total amount of cations. Then the solution is heated up to 90°C in air to promote water evaporation and to obtain a wet gel. The gel is treated up to 400°C for 2 h in air to decompose the organic framework, and the obtained powder samples were grinded and calcined at the appropriate temperature in air or in reducing atmosphere depending on the required phase.

Impregnation [3] is a process commonly used to enhance the performances of a material, and consists in depositing a new phase on an existing substrate from a solution. Wet impregnation is the more common and traditional procedure that allows depositing the required amount of active phase; where in incipient wetness, the deposited quantity is strictly related to the solubility of precursors [4]. In wet impregnation, it is sometimes difficult to reach high dispersion and is thus necessary to develop new procedures. The strategy we followed consists in depositing the precursor in complexed form so to favour the interaction with the supporting surface and limit particles coalescence.

2.2 Characterisation

The XPS measurements were carried out (with a Perkin Elmer Φ 5600ci multitechnique system) to evaluate surface composition both in terms of oxidation state of the composing elements and of quantitative analysis. At this purpose, it is interesting to compare the XPS results with EDX ones, which consider a deeper layer. Field emission scanning electron microscopy and EDX measures (around at. % accuracy) were carried on a Zeiss SUPRA 40VP. Morphological analysis and EDX analysis were carried out setting the acceleration voltages at 20 kV.

The XRD analyses were performed with a Bruker D8 Advance diffractometer with Bragg-Brentano geometry using a Cu Kα radiation (40 kV, 40 mA, λ = 0.154 nm). The data were collected at 0.03° at a counting time of 7 s/step in the (2θ) range from 20 to 70°. The crystalline phases were identified by the search-match method using the JCPDS database; the position of the peaks, furthermore, allows to evaluate, on the base of the crystal unit cell parameters, the inclusion of the dopant inside the perovskite cell. Temperature-programmed reduction (TPR) and specific surface area (BET) measurements were performed with an Autochem II 2920 Micromeritics, equipped with a TCD detector. The TPR measurements were carried out in a quartz reactor by using 50 of sample and heating from RT to 900°C at 10°C min^{-1} under a constant flow of H_2 5% in Ar (50 ml • min^{-1}). TPR samples were previously outgassed with He (50 ml • min^{-1}) at RT. The critical evaluation of the XPS and TPR data on oxidation state of cations can be very useful because TPR is a bulk quantitative analysis that considers the whole powder while XPS is surface specific one. The comparison between XPS and TPR data allows to understand segregation phenomena and to individuate the first sites responsible of the reactivity and also to evaluate the exchange and bulk intervention. In BET measurements, 100 mg of sample was used; before measurement, the sample was treated at 350°C for 2 h under a constant flow of He (50 ml • min^{-1}); each surface area obtained was the average of three consecutive measurements. Temperature-programmed desorption (TPD) measurements were performed with an Autochem II 2920 Micromeritics, equipped with a TCD detector and whose outlet was directly connected with a quadrupole mass spectrometer (QMS, ESS Genesis). The TPD measurements were carried out in a quartz reactor by using approximately 0.1 grams of perovskite powder. The data were collected under a constant flow of Helium. O_2-temperature-programmed desorption (TPD) measurements were carried out after a treatment with O_2 at 650 (for $LaCoO_3$) and 600°C (for the doped $LaCoO_3$). The O_2-TPD measurements allow to study the effect of composition and doping on the formation and stabilization of α- and β-oxygen species and thus on oxygen ions mobility and on surface and bulk oxygen vacancies.

2.3 Catalytic activity

The catalytic activity tests were carried out, at atmospheric pressure, under different conditions depending on the specific reaction and application. Measurements were carried out in a quartz reactor (6 mm ID) with a packed bed of powders (**Figure 1**); the temperature was monitored by a thermocouple right upstream of the bed reactor. The experimental set-up has been already described [5].

Soot oxidation tests were conducted both in tight and loose contact mode [6] using a 1:10 soot to catalyst ration, in a 5% O_2 atmosphere. In this case, the inert carrier used was He, and the temperature of the catalytic bed was raised from RT until no production of CO_2 could be detected.

The composition of the gas mixture (before and after reaction) was measured by GC (Agilent 7890A), with a TCD detector.

2.4 Electrocatalytic activity

Electrochemical behaviour was studied by means of impedance spectroscopy tests performed on symmetric or asymmetric cells (to evaluate the electrode or the whole device). Impedance of the cell was measured under air (normal operating

Figure 2.
Image of the experimental equipment for the test of the solid oxide cells.

condition for a SOFC cathode or SOEC anode) and in presence of fuel (H_2, CH_4). The home-made equipment is reported in **Figure 2** and is connected with an Autolab Metrohm PGSTAT204.

3. Developing functionality in perovskites for a sustainable development: results and discussion

3.1 Cases of built functionality: the art of pollutants abatement

3.1.1 Making a poorly active perovskite active in CO oxidation and NO reduction: the difficult case of $SrTiO_3$

$SrTiO_3$ is a very robust perovskite composed by very cheap, not critical and environmentally sustainable elements. Unfortunately, its catalytic activity is rather low and this compound is neither active in CO oxidation nor in NO reduction. An interesting improvement can be represented by the introduction of a redox couple in the B-site. This result can be obtained by both substituting Sr with an alkaline earth and introducing a transition metal in the B-site. Co and Mn are mostly the more active cations in oxidation; so these cations have been our first choice. In perovskite doping, the improvement of the properties (oxygen exchange and mobility, catalytic activity, etc.) derives from the insertion of dopant into the unit cell. To reach this goal, a particular care has to be devoted to evaluate the solubility and the optimal doping amount. Several attempts convinced us that a good composition could be for cobalt doping, $SrTi_{0.7}Co_{0.3}O_3$. The insertion was confirmed by the characterisation (particularly XRD and TPR). The inclusion of cobalt induces the formation of a redox couple and of oxygen vacancies that are responsible of the increased catalytic activity with respect to CO oxidation (**Table 1**). A similar attempt was carried out with Cu with the aim of increasing the catalytic activity in NO reduction. This doping, however, did not result to be of any help. In fact, no significant signal due to Cu(II) reduction is observed suggesting that once inserted into the crystalline cell, copper looses its reducibility. Usually, the reduction of copper follows its diffusion toward the perovskite surface, so we can suppose that mobility is un-favored in this case [7–12].

To take, however, advantage of the presence of copper, nanoparticles of CuO were deposited onto the surface of $SrTi_{0.7}Co_{0.3}O_3$ by wet impregnation. This allowed to add the reactivity towards NO reduction due to the copper nanoparticles and to the reactivity towards CO oxidation due to Co-doped perovskite.

	$SrTi_{0.7}Co_{0.3}O_3$		$CuO/SrTi_{0.7}Co_{0.3}O_3$	
	CO oxidation	CO + NO	CO oxidation	CO + NO
Temperature °C	CO Conversion	NO conversion	CO Conversion	NO conversion
100	0	0	0	0
200	19	0	2	8
300	84	0	43	43
350	91	0	68	100
400	90	0	79	100

Table 1.
Conversion values obtained, at different temperature, on the Co-doped $SrTiO_3$ and in the nanocomposite $CuO/SrTi_{0.7}Co_{0.3}O_3$ obtained by wet impregnation.

In the nanocomposite, the activity in CO assisted NO reduction greatly increases (**Table 1**) while that in CO oxidation does not decrease significantly.

$SrTiO_3$ was also activated with a different strategy consisting in doping both in Sr- and Ti-sites. For Sr-site, the substitution with K was attempted, whereas for the Ti-site, Mn was preferred. We expected that both oxygen vacancies formation and Mn-based redox couples could form and be precious in enhancing the catalytic activity [13]. This hypothesis was confirmed by the experimental data (**Table 2**). In the undoped $SrTiO_3$, no activity was observed. When Mn is added, the conversion in CO oxidation increases from 18 to 94% at 600°C. The best performance for CO assisted NO reduction is observed when a composite is obtained by depositing CuO on K and Mn-doped $SrTiO_3$.

Also the soot oxidation is improved by doping in the Sr-site and Ti-site (**Table 3**).

3.1.2 Copper activation: inside the perovskite cell or deposited onto the surface?

The behaviour of dopants and the capability to improve the reactivity depend on the specific perovskite and can be discussed in terms of structural and chemical stabilization. The insertion of copper as a substitute of cobalt in a $LaCoO_3$ perovskite, as an example, greatly increases the reactivity of undoped perovskite with

	Reaction	Temperature (°C)	Conversion CO	Conversion NO	Conversion CO
SKT	$CO + O_2$	400	0		
	$CO + O_2$	600	18		
	CO + NO	400		2	5
	CO + NO	600		5	10
$SKTM_{30}$	$CO + O_2$	400	54		
	$CO + O_2$	600	94		
	CO + NO	400		25	25
	CO + NO	600		50	49
$Cu/SKTM_{30}$	CO + NO	400		100	94
	CO + NO	600		100	100

Table 2.
Conversion values obtained at 400 end 600°C for the K-doped, K and Mn-doped $SrTiO_3$ ($Sr_{0.8}K_{0.2}TiO_3$, SKT—$Sr_{0.8}K_{0.2}Ti_{0.7}MN_{0.3}O_3$—$SKTM_{30}$) and for a nanocomposite obtained by depositing CuO on Mn-doped $SrTiO_3$ ($Cu/SKTM_{30}$).

Catalyst	T_{onset}	T_{max}	CO_2 conversion
SKT	307	346	89
$SKTM_{15}$	276	333	89
$SKTM_{30}$	261	329	89
$SKTM_{30}$ 800	264	327	81

Table 3.
Onset temperature and temperature of maximum conversion of soot in presence of 10% O_2 and with a soot: catalyst of 1:10. The amount of converted CO_2 with respect to the total expected CO_2 is also reported ($Sr_{0.8}K_{0.2}TiO_3$, SKT—$Sro.8K_{0.2}Ti_{0.85}Mn_{0.15}O_3$—$SKTM_{15}$; $Sr_{0.8}K_{0.2}Ti_{0.7}Mn_{0.3}O_3$—$SKTM_{30}$). All samples are treated at 700°C calcination temperature; the effect of increasing the calcination temperature is reported comparing $SKTM_{30}$ (treated at 700°C) and $SKTM_{30}$ 800 (treated at 800°C).

respect to the NO reduction. Several aspects contribute to this result. The first copper insertion into the lattice causes the formation of oxygen vacancies for electroneutrality reasons. The NO molecules interact dissociatively with the surface oxygen vacancies that, in this way, play a key role in NO activation (usually the rate determining step). CO can interact with the Lewis active sites distributed on the surface (Co cations, as an example), which is the first step for its oxidation [14, 15].

Doping can also increase the oxygen mobility and exchange capability, which is fundamental for oxidation, considering the Mars-van-Krevelen mechanism; in fact, the capability of perovskite to oxidize CO is fundamental.

The effect of doping with Cu was investigated by our research group and compared with the case of Cu deposited on $LaCoO_3$ (**Tables 4** and **5**, **Figure 3**).

Copper inside the perovskite cell, as mentioned, increases the reactivity thanks to the formation of oxygen vacancies activating small molecules but under oxidizing environment, it is difficult for oxygen vacancies to survive on surface and thus the activating effect of copper is due to the capability of this cation to coordinate more than one NO molecule facilitating the formation of the N_2O intermediate that evolves into the formation of N_2 and O_2.

At high Cu/Co atomic ratios, there is no significant difference between the two synthetic approaches, but with lower Cu contents, the nanocomposites seem more active in CO oxidation. A complex situation is observed in NO reduction: at lower Cu/Co atomic ratios, the $LaCo_{1-x}Cu_xO_3$ catalysts are more active than the nanocomposites. This suggests that the surface is mainly involved in the NO

Temperature	Reaction	x = 0	x = 10	x = 15	x = 20	x = 30	x = 15 WI
200	$CO + O_2$	80	7	75	69	73	4
	CO+ NO	0	2 (1)	0 (0)	0 (0)	0 (0)	0 (0)
250	$CO + O_2$	92	87	88	85	90	73
	CO+ NO	0	6 (3)	4 (5)	4 (0)	0 (0)	0 (0)
300	$CO + O_2$	94	91	93	92	93	85
	CO+ NO	0	18 (17)	62 (56)	67 (59)	42 (36)	6 (5)
350	$CO + O_2$	96	94	94	94	97	94
	CO+ NO	2 (1)	39 (36)	80 (71)	81 (66)	94 (76)	68 (63)
400	$CO + O_2$	100	100	100	100	100	100
	CO+ NO	4 (2)	100 (92)	100 (92)	97 (92)	100 (92)	83 (77)

Table 4.
Conversion values obtained, at different temperature, on the xCuO/$LaCoO_3$ with x = atomic percentage.

Temperature	Reaction	0	0.1	0.3	0.5
200	$CO + O_2$	80	5	2	94
	CO+ NO	0	0 (3)	13 (0)	0 (2)
250	$CO + O_2$	92	24	46	93
	CO+ NO	0	8 (2)	11 (23)	45 (42)
300	$CO + O_2$	94	66	67	96
	CO+ NO	0	8 (7)	24 (25)	63 (58)
350	$CO + O_2$	96	86	79	96
	CO+ NO	2 (1)	52 (45)	80 (72)	83 (70)
400	$CO + O_2$	100	94	95	96
	CO+ NO	4 (2)	71 (77)	100 (95)	97 (95)

Table 5.
Conversion values obtained, at different temperature, on the doped perovskite $LaCo_{1-x}Cu_xO_3$.

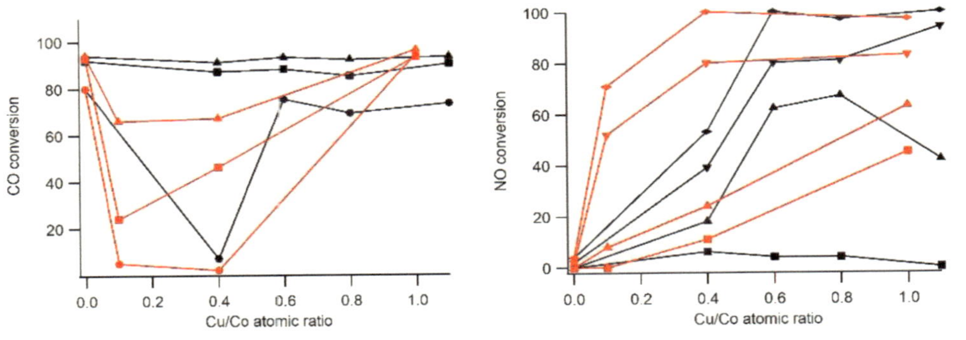

Figure 3.
(Left) CO conversion (in CO oxidation) and (right) NO conversion (in CO + NO) observed as a function of the Cu/Co atomic ratio for the catalysts with copper inside the cell (red) and deposited on the $LaCoO_3$ surface (black). ● 200°C, ■ 250°C, ▲ 300°C, ▼ 350°C and ✦ 400°C.

activation by means of oxygen vacancies until Cu content is high enough to favour another mechanism, frequently observed in simple oxides that, unlike perovskites, are not characterised by the presence of oxygen vacancies. Following this mechanism, NO molecules are activated by coordination to the Cu clusters.

Other nanocomposites can behave in a different way: in the soot oxidation carried out with $LaMnO_3$-based perovskites, as an example. To enhance the activity in oxidation, we doped substituting Mn with Co. Co oxides are very active catalyst for oxidation [16] but the use of cobalt is not very easy from an industrial point of view. So we decided to use this element as a dopant of Mn. In this case, we used K as a dopant in the A-site in the attempt to oxidize soot comparing with the effect of depositing K oxide on the surface of the perovskite. It is evident from the temperature of maximum conversion (**Table 6**) that the better result (lower maximum activity temperature, T_{max}) is obtained when La is used as a 12-coordinated cation (pushing manganese to be Mn(III)) and K is inserted into the perovskite unit cell.

Also the doping in the A-site is a powerful mean to increase the reactivity [17]. In this case, the aliovalent substitution of La with an alkali earth or an alkaline element determines both the formation of the oxygen vacancies and the increment of the oxidation state of the cation in the B-site. As an example, in the $LaCoO_3$, the substitution with Sr can favour also the formation of Co in unusual oxidation state as IV but the low stability of this specie suggests also the formation of oxygen vacancies. The equilibrium between these two effects of doping depends on the redox potential, the strength of the M-O bonds, etc. allowing for the creation of several different and complementary reaction mechanisms whose relevance depends also on temperature. In $LaCoO_3$-based catalysts, as an example, around 350°C, the CO oxidation mechanism changes from suprafacial to intrafacial; pulsed

Sample	T_{max}
Only soot—no catalyst	627
$LaMn_{0.9}Co_{0.1}O_3$	330
$La_{0.9}K_{0.1}Mn_{0.9}Co_{0.1}O_3$	306
$Sr_{0.9}K_{0.1}Mn_{0.9}Co_{0.1}O_3$	323
$0.1\ K/LaMn_{0.9}Co_{0.1}O_3$	330

Table 6.
Temperature of maximum activity in soot oxidation (10% O_2 + 5000 ppm NO) for different $LaMnO_3$-based catalysts.

Sample	Reaction	200°C	250°C	300°C	350°C	400 °C
$LaCoO_3$	MSR	0	0	0	0	0
	ESR	0	0	0	0	7
	MOSR	2	15	28	13	24
	EOSR	1	11	12	41	37
$LaCo_{0.7}Cu_{0.3}O_3$	MSR	0	30	50	67	80
	ESR	4	24	28	20	24
	MOSR	0	6	51	84	90
	EOSR	1	12	43	70	77
$La_{0.7}Sr_{0.3}CuO_3$	MSR	0	3	12	18	14
	ESR	0	7	48	68	35
	MOSR	0	7	9	99	100
	EOSR	0	3	8	21	100

Table 7.
Conversion obtained in un-doped and doped $LaCoO_3$ in methanol steam reforming (MSR), ethanol steam reforming (ESR), methanol oxidative steam reforming (MOSR) and ethanol oxidative steam reforming (EOSR).

experiments optimised for perovskites allowed to evaluate that the activation energy decreases from 40 (for undoped) to 29 (Sr-doped) to 27 (Cu-doped) kJ/mol [18]. Doping can have a different effect on the activity and selectivity being, these properties, related to different properties of the catalysts. The substitution of La with Sr increases the activity in NO reduction, when a complex mixture simulating the automotive exhaust is used, but not for the CO assisted NO reduction [17].

In $LaMnO_3$, the formation of oxygen vacancies is, in contrast, less favoured with respect to that of Mn(IV) which is a stable state for manganese; so the catalytic properties can be tuned in a different way.

So depending on the required catalytic activity, different dopants and different reactions can be enhanced. A similar strategy can be successfully applied to processes different from abatement of pollutants. A lot of relevance, as an example, is gathered by the steam reforming and oxidative steam reforming of alcohol. Also in this case, the doping with copper greatly improves the activity of $LaCoO_3$ both in ethanol and even more in methanol steam reforming and oxidative steam reforming. The results are surely due to the presence of copper, as demonstrated by the activity and selectivity of a $LaSrCuO_3$ in methanol and ethanol steam reforming and oxidative steam reforming (**Table 7**) [19, 20].

3.1.3 NOx reduction: a very difficult challenge. Improvements derived from smart nanocomposition

The reactivity developed in perovskite-based nanocomposites is strongly dependent on the supporting perovskite and on the interaction of the surface with the deposited phase. In particular, it is widely accepted that higher performance can be obtained through high dispersion of activating particles. With the aim of a deeper understanding, traditional wet impregnation and an innovative ammonia driven precipitation procedure were compared in terms of dispersion (by means of TPR) and catalytic activity. Both model reactions (CO oxidation and CO assisted NO reduction) and reactions with a synthetic automotive exhaust mixture, including 10% steam, and oxygen, were carried out. The catalytic results clearly underline that higher dispersion means higher catalytic activity and that the difference is particularly evident at lower temperature. To favour the dispersion of nanoparticles, the interaction at the interface is paramount. A strong interaction prevents the particles sintering and allows to expose a higher amount of active sites.

It has to be considered that some molecules are activated at the interface steps where coordinatively unsaturated sites are more accessible [15, 21, 22].

A good choice of the support (that can be characterised by a specific reactivity) and of the activating nanoparticles allows to build an interesting reactivity. In our work, we compared the deposition of copper oxide carried out on $LaCoO_3$ and $LaNiO_3$. In both cases, the nanocomposites show good activity in oxidation and in NO reduction, with performance greatly enhanced with respect to the single components, but in the $CuO/LaNiO_3$ case, the activity in NO reduction is increased thanks to the presence of H_2 that originates from the steam reforming catalyzed by the support; so, a successful synergic approach is obtained.

We compared the results with the ones obtained in similar reactions with $CuO/LaCoO_3$. Different interactions and synergies were observed with respect to $CuO/La_{0.5}Sr_{0.5}CoO_3$. Sr-doping, in fact, enhances oxygen mobility affecting the reducing character of the nanodispersed CuO and thus the reactivity under different conditions [23].

The deposition of copper oxide significantly increases the activity of the nanocomposites in CO oxidation (about 100% conversion at 200°C) and in CO + NO (50% conversion at 250°C, more than 80% at 400°C) reactions. Comparing the nanocomposites obtained by depositing CuO clusters in undoped and Sr-doped lanthanum cobaltate, it is evident that the main difference concerns the compounds with lower Cu content. The catalysts obtained using the Sr-doped perovskite became highly active at lower temperature. On simulated gasoline engine exhaust, the nanocomposites always improve the oxidation activity compared to the parent perovskite, while the NO reduction is quantitative in the absence of O_2.

3.1.4 From the powder to a functional device: the art of deposition

Once the catalytic perovskite is developed and optimised from the compositional, morphological and structural point of view, the device realization is the next challenge. An interesting example is the deposition of the active layer onto a cordierite support, which is usually carried out in three-way catalysts systems. The traditional approach is rather time- and material-consuming, being composed by several steps of immersion in a perovskite suspension followed by heat treatments carried out slowly enough to avoid cracks in the active layer. With the aim of saving time and materials and thus to increase the sustainability of the catalytic system, we developed, starting from the Marcilly wet chemistry synthesis, an innovative procedure consisting in growing the perovskite directly on the cordierite honeycomb [24].

The cordierite support was immersed into a solution of precursors also containing the citric acid. A pH around 6–7 was obtained by adding drop to drop an aqueous solution of ammonia. The wet honeycomb was thus treated at the calcination temperature which is necessary to obtain the perovskite structure. In this way, the perovskite was grown directly on the support in a single step. Beside the time-saving, also the catalyst is required in lower amount and the cordierite surface is completely covered by the active perovskite. To evaluate this, the La/Si XPS atomic ratios as a function of weight increment have been reported (**Figure 4**). Fadley [25, 26] and Argile et al. [27], in fact, have shown that the atomic ratio between the XPS signals of supported and supporting species changes with the deposited amount in a different way for a homogeneous or a 'island' growth mechanism. The comparison with the literature data indicates, for the direct method, a deposition of a homogeneous layer and, for the washcoating procedure, the deposition by islands. This behaviour is also confirmed by SEM images (**Figures 5** and **6**). Moreover, the amount of active layer deposited by direct procedure can be tuned by tuning the composition of the precursors' solution.

Finally, the porosity generated by gas evolution during the synthesis (comparing **Figures 5** and **6**) allows to reach a good surface area that contributes to increase the catalytic performance.

3.2 Cases of built functionality: carbon dioxide; from green-house-gas to synthetic fuel; and improvement of catalytic activity through exosolution

Methane dry reforming (MDR) is an endothermic reaction of high scientific and industrial importance that requires a very high temperature [28–31].

In spite of being studied from 1888, it is not yet considered an industrially mature process, mainly because of the C-poisoning and sintering of the catalysts which are based on noble metals or nickel. Several aspects affect the performances of catalyst: the strong interaction between supporting and active phases, the dispersion and size of active phase, the basicity, the reducibility, the oxygen exchange capability, porosity and specific surface area [29].

The mechanism of methane dry reforming has been investigated on several catalysts [32–34]. In general, four steps are considered: (1) dissociative adsorption of methane—which is the rate determining step and should be favoured by step active sites; (2) dissociative adsorption of CO_2—which is generally considered fast particularly on metal-support interface; (3) hydroxyl groups formation; and (4)

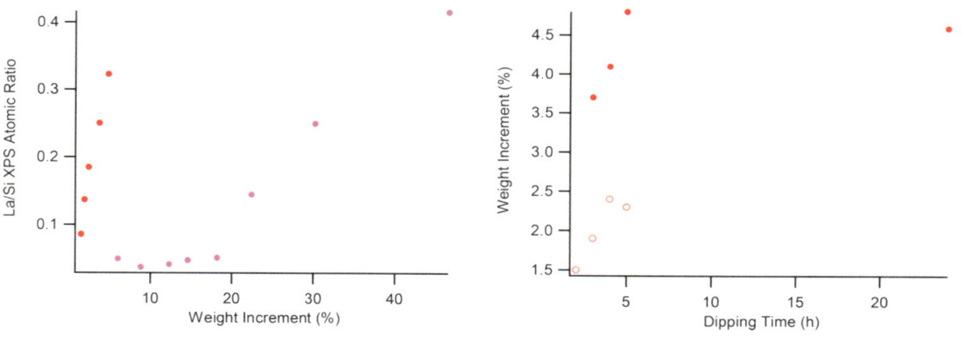

Figure 4.
La/Si atomic ratio (corresponding to the degree of coverage of the cordierite by $LaCoO_3$, being XPS a surface specific technique) observed in the direct (red) and washcoating (magenta) deposition procedures (left). Amount of catalyst deposited (evaluated by weight increment) with a solution of 12 g/l (filled circles) and 1.2 g/l (empty circles) of precursor (right).

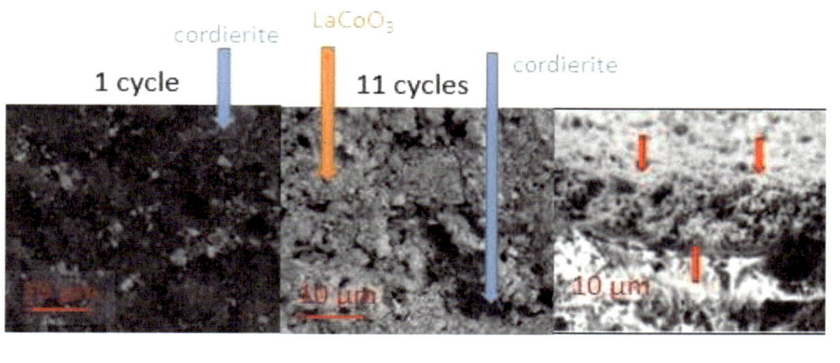

Figure 5.
SEM images of the cordierite surface during the increasing washcoating cycles and of the section.

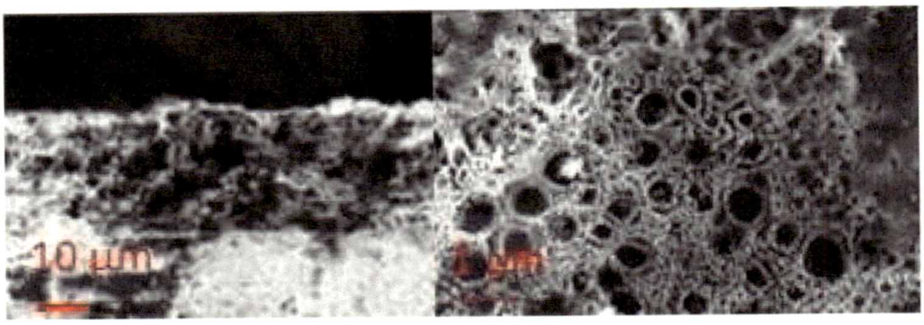

Figure 6.
SEM images of the cordierite/$LaCoO_3$ section and surface after the deposition carried out by direct deposition.

intermediates oxidation and desorption—surface oxygen is considered responsible of the conversion of CH_x –groups adsorbed on the surface in CO and H_2. There is no consensus on the detailed reactions composing this last step but this can become a slow step depending on the specific support and catalyst.

Perovskites are frequently considered as a good strategy for improving the catalytic performance in MDR. In fact, there is a powerful technique capable of allowing the creation of highly dispersed and stabilized Ni clusters on the surface: exosolution [35]. Exosolution consists of a treatment of the perovskite (in this case, a Ni-containing perovskite) under reducing condition. This induces the Ni to exit from the perovskite cell and to migrate towards the surface creating a strong interaction with it. This strong interaction provides higher stability towards the Ni migration and sintering during reaction. Moreover, if lanthanum is present in the perovskite, a lanthanum oxycarbonate specie forms, preventing the C-poison.

To test the effect of copper doping into the activity of $LaNiO_3$-based catalysts in MDR, we compared $LaNiO_3$ and $LaNi_{0.7}Cu_{0.3}O_3$. Moreover, the nickelate has been modified by exosolution and by deposition of NiO nanoparticles.

The catalysts have been prepared by citrate route; the XRD patterns indicate that the crystalline phase is obtained after thermal treatment at 850°C; the BET specific surface area is 10–11 m^2/g and is not affected by the insertion of copper.

Also the morphology is not really modified by the presence of copper (**Figure 7**).

The H_2 consumption determined in the TPR analysis indicates that in the $LaNiO_3$ sample, 17% of Ni is present as Ni(III), whereas in the Cu-doped, the % increases to 46%. This difference suggests that Cu increases the reducibility of Ni in the perovskite. This phenomenon, already observed for other perovskites, is attributed, following Tien-Taho et al. [36, 37], to the attitude of copper, under reducing conditions, to migrate towards the surface. Elemental copper distributed on the surface catalyses the H_2 activation decreasing the reduction temperatures. In fact, two reduction peaks are observed in the TPR curve of LaNiO3: one is centred at 434°C and is attributed to the Ni(III) to Ni(II) process, and the other at 595°C is due to the Ni(II) to Ni(0) reduction. When copper is inserted into the structure, the first peak (now corresponding to Ni(III) to Ni(II) and to Cu(II) to Cu(0) reductions) is shifted at 333°C and the second one at 566°C. XPS analysis allows to add some complementary and interesting information. Beside the peak shape and position, consistent with the presence of La(III) (see also the shake-up signal at 838 eV characteristic of this oxidation state for La—**Figure 8**) and Cu(II) (shake-up signal at 942–944 eV—**Figure 8**), it is evident that O 1 s signal (**Figure 8**) is composed by two contributions attributed to perovskite lattice oxygen (528.8–529.0 eV) and to hydroxyl groups (531.7 eV). The quantitative analysis (**Table 8**) confirms the presence of surface contaminants (hydroxyl groups): the atomic percentage of oxygen is higher than the nominal one, particularly in the undoped sample. Lanthanum is surface segregated in

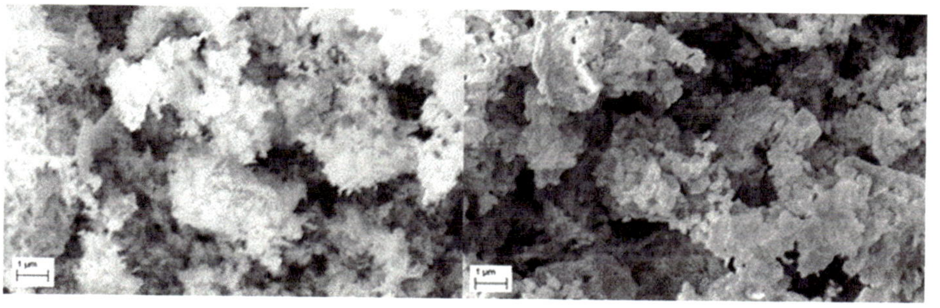

Figure 7.
SEM images obtained for the undoped (left) and Cu-doped (right) $LaNiO_3$.

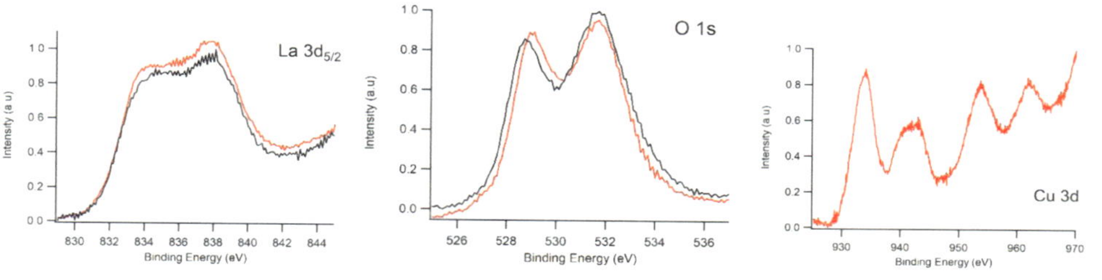

Figure 8.
XP spectra obtained for the undoped (black) and Cu-doped (red) $LaNiO_3$.

Sample		La	Ni	Cu	O	O/(La + Ni + Cu)	Cu/Ni
$LaNiO_3$	XPS	18.4	7.6		74.0	2.85	0.00
	Nominal	70.9	29.1		60.0	1.50	0.00
		20.0	20.0				
		50.0	50.0				
$LaNi_{0.7}Cu_{0.3}O_3$	XPS	15.3	4.6	13.8	66.3	1.97	3.00
	Nominal	45.5	13.6	40.9	60.0	1.50	0.43
		20.0	14.0	6.0			
		50.0	35.0	15.0			

Table 8.
Atomic composition obtained with XPS for the undoped and Cu-doped $LaNiO_3$ compared with the nominal one determined by the weighted amounts.

the undoped sample but not in the doped one. In this case, it is evident that the copper is surface segregated (and this is expected to affect reactivity).

The reactivity of these doped and undoped nickelates is summarized in **Figure 9**.

No particularly evident differences can be observed between undoped and doped catalysts in spite of the relatively high amount of Cu inserted. A marked difference, by contrast, is evident by comparing the activity before and after the exosolution pretreatment (**Figure 10**). In this last case, in fact, the catalyst starts its activity at 600°C instead of 800°C.

3.3 Cases of built functionality: from catalysis to electrocatalysis; solid oxide fuel cells and solid oxide electrolysis cells

Beside catalyst-based systems, very important devices focusing on sustainable development are solid oxide fuel cells (SOFCs) and solid oxide electrolysers (SOECs) [38, 39].

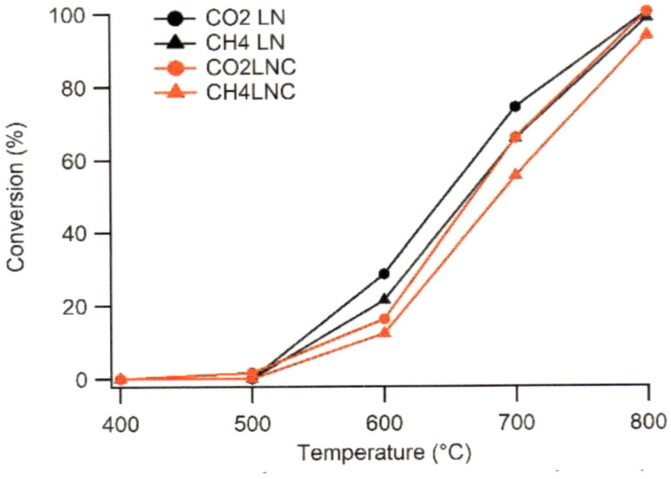

Figure 9.
CO_2 and CH_4 conversion values obtained with undoped $LaNiO_3$ (LN) and Cu-doped one (LNC).

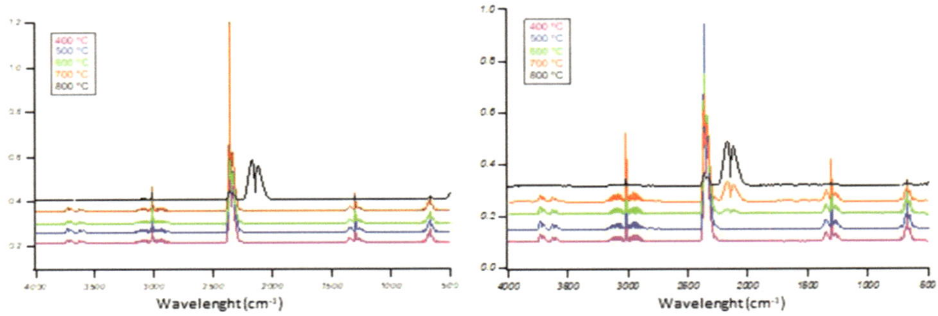

Figure 10.
IR spectra obtained, as a function of temperature, upon exposure of the as-prepared (left) and exosoluted (right) $LaNiO_3$ to CO_2 + CH_4.

SOFCs are electrochemical devices capable of converting the chemical energy of fuels into electrical energy; *vice versa* SOECs are devices that can convert the electric energy deriving, in excess, from intermittent renewable sources into the chemical energy of a fuel. Unlike other fuel cells, SOFCs operate at intermediate to high temperature (700–1000°C), and consequently, no noble metals are necessary to activate the electrodes. Another advantage is offered by the possibility to reverse the cell, that is, making it work as a fuel cell or as an electrolyser. This approach could be very advantageous because the endothermic requirements can be fulfilled by electric energy and temperature; moreover, the reversibility allows to minimise the problems related to the C-poisoning of the surface (when fuels different than pure hydrogen are used) and the electrode/electrolyte delamination. However, robust and durable materials are necessary due to the severe working conditions and, if reversible SOFCs are desired, the stability and catalytic activity are required both under oxidizing and reducing conditions. In our research group, we aim at developing reversible intermediate temperature (600–700°C) solid oxide fuel cells operating with C-containing fuels.

To reach this objective, we considered, at first, the starting point: (La, Sr)(Fe, Ga)O_3 (LSGF). LSGF has only slightly lower performances with respect to the benchmark perovskite-based electrodes, but it exhibits an excellent chemical stability.

Strontium and gallium insertion enhances the oxygen mobility; in addition, gallium seems to enhance chemical stability in reducing conditions. Iron contributes to the electrochemical performances and also increases the electronic conductivity of the materials. Mixed ionic and electronic conductivity (MIEC) is a particularly important property for the electrodes in SOFCs. In fact, unlike traditional electronic conductors, MIEC allows to extend the active surface avoiding the problem of triple phase boundary (TPB). The composition exhibiting the best compromise between activity and stability has been identified in $La_{0.6}Sr_{0.4}Ga_{0.3}Fe_{0.7}O_3$; hence, this is the starting point.

We have successfully synthesized $La_{0.6}Sr_{0.4}Ga_{0.3}Fe_{0.7}O_3$ with a simple, controllable and easily scalable wet chemistry route: preparation with amounts from 0.5 to 15 g has been carried out and the obtained results demonstrated the high reproducibility of the method. We have investigated several parameters of the synthesis, setting the minimum amount of nitric acid at 4.5 ml per gram of product for a complete combustion, and the minimum calcination temperature at 900°C for a completely pure phase. The calcination temperature limits the superficial area, which is useful in particular for catalytic purposes, to 9 m^2/g. Calcination temperature can be lowered if minor impurities are tolerated or whether the material will undergo cycles of reduction/oxidation. In this case, the perovskite auto-arranges towards the complete purity and the BET specific surface area is about twice. Investigations on the powders prior to calcination indicated that after the combustion, the perovskitic phase has already been formed, but persists a massive organic fraction that requires a temperature treatment for its removal. The resistance of the material to the reduction was tested between 800 and 1000°C, and the obtained results outline an exceptional resistance to reduction compared to the other perovskites: operation of the material in hydrogen atmospheres at 800°C leads only to minor structural modification. The resistance to reduction and the capability to recover the original structure after an oxidation treatment make this perovskite particularly suitable also for applications requiring tolerance to very reducing atmosphere. No differences of stability have been observed as a function of the different preparation conditions.

Once defined the starting point we considered necessary to improve the electrocatalytic activity. Also for this application, wet impregnation for nanocomposition was fundamental. To minimise the addition of complexity, we decided to create a nanocomposite of the type FeOx/LSGF for the cathode.

The use of nanocomposites powders as starting point for electrodes allows to deeply modify the electrochemical performance. A thin, Sr/Fe-rich foil forms on the surface of the electrode during the SOFC thermal treatment and deeply improves the electrochemical behaviour of the FeO_x/LSGF cathode. The electrochemical results are encouraging for future application in SOFCs, as nanocomposite has an ASR of 2.1 $\Omega \cdot cm^2$ at 620°C, only ⅓ of LSGF's one in the same conditions.

3.4 Cases of built functionality: oxygen permeable dense ceramic membranes

The efficiency of a fuel cell depends on (among other elements) the internal losses in the cell, including the ohmic resistance of the electrolyte and the overpotential losses at the electrodes. While the first one is well understood today, the second one needs a deeper characterisation. At the cathode site, in particular, the oxygen reduction is generally thought to be the most difficult reaction to activate in a SOFC. The mechanism appears rather elaborate and various aspects need to be considered [40, 41].

First, molecular O_2 needs to be converted into some 'electro-active' intermediate form via one or more processes. These reactions do not depend on the current (except in the limit of steady state) and they are driven by the chemical potential (depletion or surplus of the intermediate). The second important point concerns the diffusion of the intermediate species (mass transfer) through the electrode (to the electrolyte). Since electrochemical reactions and diffusion occur cooperatively over an active area, the overall rate is co-limited by both these processes.

The oxygen permeation is a selective phenomenon that allows separating oxygen from a mixture of gases. **Figure 11** schematizes the permeation mechanism.

Oxygen undergoes reduction at the O_2-rich side, and then the oxide ions are transported through the membrane at the O_2-deficient side. The last step involves the re-oxidation of the oxide ions and the release of molecular oxygen. As can be seen, the driving force of the overall process is the O_2 pressure gradient between the two sides of the membrane.

The permeation mechanism involves both the electrochemical and the transport properties of the investigated material.

It is now opportune to point out that the assumed mechanism can only be carried out by a MIEC. In fact, electrons must be able to move to the O_2-rich side to reduce O_2, and the oxide ions (generated at the oxygen-rich side) need to reach the O_2-deficient side.

The last feature involved in the permeation mechanism concerns the chemical properties of the material with respect to the reduction and oxidation of oxygen. From the statements above, a strong relationship between oxygen permeation properties and cathode activity is well evident: the processes involved in oxygen permeation and in SOFCs cathode working are very similar and the only difference concerns the last step (i.e. re-oxidation of oxide ions in the permeation process, compared to oxide ions transfer to the electrolyte in the SOFCs).

It should be evident that a high oxygen permeation rate can guarantee an overall fast operation when the material is employed as cathode in SOFCs.

The permeation phenomena can be described taking into account the oxygen ions and the electrons diffusion through the membrane. In detail, the working scheme summarized in **Figure 11** (left) suggests that electrons and ions move in opposite directions in order to give a total current density equal to zero under steady state conditions. Under these conditions, the oxygen flow, JO_2, is strictly bound to the MIEC.

Doping can really increase the oxygen permeation both increasing the oxygen vacancies and mobility and surface reactivity, as evidenced by the JO_2 values obtained on $LaCoO_3$ undoped and doped with Sr and with Cu (**Table 9**).

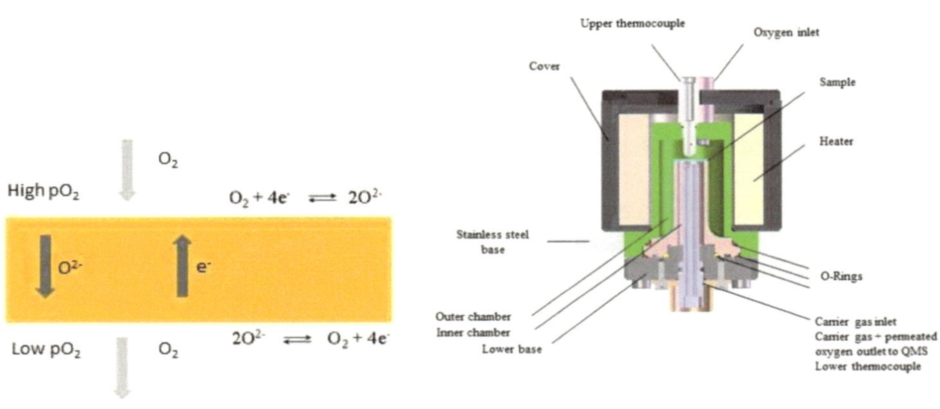

Figure 11.
Schematic view of the oxygen permeation in a mixed electronic-ionic conductor (left) and of the device (right).

Sample	500	600	700
$LaCoO_3$	0.7	0.7	0.8
$LaCo_{0.5}Cu_{0.5}O_3$	1.6	1.5	1.6
$La_{0.5}Sr_{0.5}Co_{0.5}Cu_{0.5}O_3$	0.8	1.0	1.3

Table 9.
O_2 flow, $JO_2 \times 10^{-2}$ mL/cm² min obtained as a function of temperature in undoped and doped $LaCoO_3$.

4. Conclusions

Perovskites are versatile oxides whose properties can be tuned finely through different approaches. In this case, our focus is on ABO_3 perovskites. Among the relevant properties of oxide perovskites, we have considered catalysis and electrocatalysis. Both these fields are very interesting for sustainable development. Catalysis is relevant for abatement of pollutants (NOx, TWC, as an example) or for the production of green hydrogen and the conversion of carbon dioxide into fuel (dry reforming of methane); electrocatalysis is the basis for the realization of solid oxide cells (fuel cells and electrolysers). Several aspects are fundamental both for catalysis and electrocatalysis, such as the presence of active sites on the surface of the perovskites or the development of oxygen mobility and exchange.

To develop these properties in perovskites, there are several possibilities: in this contribution, we underline the opportunities offered by doping, nanocomposition and exosolution.

Several perovskites of the type $LaCoO_3$, $SrTiO_3$, $LaMnO_3$ and $LaFeO_3$ have been considered and the effect of doping in the A-site (Sr, K) and B-site (Cu, Mn, Co) has been evaluated.

When a complex functionality is required, nanocomposition is sometimes preferred. Nanocomposition can be carried out by different procedures. In this contribution, the materials have been obtained by wet impregnation, a traditional procedure, and by ammonia driven precipitation, an innovative method that, based on the use of complexants, allows obtaining highly dispersed particles. Several composites of the type CuOx/Perovskite ($LaCoO_3$, $(La,Sr)CoO_3$, $LaNiO_3$ and $SrTi_{0.7}Co_{0.3}O_3$) have been investigated and compared. The catalytic activity depends on the interaction between the deposited active phase and the support, and on the exchange properties of the system.

A further complexity characterises electrocatalysts: beside the catalytic attitude, the conductibility is fundamental. In our research, the focus is on the development of electrodes for SOFCs and SOECs (the fuel and air electrodes). To this purpose, an important improvement is constituted by the simultaneous presence of ionic and electronic conductivity (mixed ionic electronic conductivity—MIEC). In MIEC electrodes, the reactions extend over the whole surface, whereas in the purely electronic conductors, it is possible only at the interphase between electrode/electrolyte reached by gas phase reactants (triple boundary zone). One of the bottlenecks to the commercialization of solid oxide cell is due to the cost and durability. Both these aspects are connected to the high-operating temperature (800–1000°C). So, beside MIEC, it is very important to decrease the operating temperature approaching intermediate temperature. Finally, to go further, it is useful to be able to use C-containing fuels that can be derived from biogas, natural gas, etc. In our research, we developed several electrodes active at 600–700°C in solid oxide fuel cells and solid oxide electrolysis cells. To assure stability, we also considered very robust perovskites of the type $SrTiO_3$ or $La_{1-x}Sr_xGa_{1-y}Fe_yO_3$, increasing their performance by doping, nanocomposition and infiltration.

By means of these strategies, we obtained performant electrodes and, moreover, in some cases, electrodes for reversible solid oxide cells. Reversible cells can operate both as SOFCs and as SOECs and, finally, can behave as super-batteries for energy storage.

The oxygen mobility added to the activity in oxidation and in reduction can make perovskites good candidate for the realization of dense ceramic membranes for the separation of gases. We studied, as an example, the effect of doping on the permeability to O_2 of $LaCoO_3$-based perovskites.

In this contribution, our aim was in demonstrating that, by means of doping, nanocomposition and exosolution, we can create highly active catalysts and electrocatalysts. This was also shown considering the dry reforming of methane, a very important reaction capable of converting greenhouse gases into fuel (syngas). Without the use of noble or costly metals, we developed sustainable doped $SrTiO_3$ active in DRM at 650–700°C.

Also the conversion of a powder into device has been considered focusing on the case of three-way catalysts. An innovative procedure for the 'deposition' of the active perovskite phase on cordierite was developed and the obtained results have been compared to those obtained by means of the traditional washcoating procedure: both catalysts and time waste are avoided and an efficient system is obtained by direct deposition.

Acknowledgements

The research leading to these results has received funding from the European Union's H2020 Programme under grant agreement 686086 PARTIAL-PGMs.

Conflict of interest

The authors declare no conflict of interest.

Author details

Andrea Bedon[1] and Antonella Glisenti[2]*

1 Efesto Innovation, Padova, Italy

2 Department of Chemical Sciences, University of Padova, Italy

*Address all correspondence to: antonella.glisenti@unipd.it

References

[1] Peña MA, Fierro JLG. Chemical structures and performance of perovskite oxides. Chemical Reviews. 2001;**101**: 1981-2017. DOI: 10.1021/cr980129f

[2] Marcilly C, Courty P, Delmon B. Journal of the American Ceramic Society. 1970;**53**:56-57. DOI: 10.1111/j.1151-2916.1970.tb12003.x

[3] Schwarz JA, Contescu C, Contescu A. Methods for preparation of catalytic materials. Chemical Reviews. 1995;**95**: 477-510. DOI: 10.1021/cr00035a002

[4] Kung HH. Transition Metal Oxides: Surface Chemistry and Catalysis – Studies in Surface Science and Catalysis. Vol. 45. Elsevier Science Publishers; 1989

[5] Guiotto M, Pacella M, Perin G, Iovino A, Michelon N, Natile MM, et al. Applied Catalysis A: General. 2015;**499**: 146-157. DOI: 10.1016/j.apcata.2015.04.013

[6] Mul G, Neeft JPA, Kapteijn F, Makkee M, Moulijn JA. Applied Catalysis B: Environmental. 1995;**6**: 339-352

[7] Glisenti A, Natile MM, Carlotto S, Vittadini A. Co- and Cu-doped titanates: Toward a new generation of catalytic converters. Catalysis Letters. 2014;**144**: 1466-1471. DOI: 10.1007/s10562-014-1294-5

[8] Carlotto S, Natile MM, Glisenti A, Vittadini A. Electronic structure of $SrTi_{1-x}M_xO_3$-δ (M = Co, Ni, Cu) perovskite-type doped titantate crystals by DFT and DFT plus U calculations. Chemical Physics Letters. 2013;**588**: 102-108. DOI: 10.1016/j.cplett.2013.10.020

[9] Carlotto S, Natile MM, Glisenti A, Vittadini A. Adsorption of small molecules at the cobalt-doped $SrTiO_3$(001) surface: A first-principles investigation. Surface Science. 2015;**633**: 68-76. DOI: 10.1016/j.susc.2014.11.025

[10] Carlotto S, Natile MM, Glisenti A, Vittadini A. Adsorption of CO and formation of carbonates at steps of pure and Codoped $SrTiO_3$ surfaces by DFT calculations. Applied Surface Science. 2016;**364**:522-527. DOI: 10.1016/j.apsusc.2015.12.194

[11] Carlotto S, Natile MM, Glisenti A, Paul J-F, Blanck D, Vittadini A. Energetics of CO oxidation on lanthanide-free perovskite systems: The case of Co-doped $SrTiO_3$. Physical Chemistry Chemical Physics. 2016;**18**: 33282-33286. DOI: 10.1039/c6cp03994d

[12] Carlotto S, Natile MM, Glisenti A, Vittadini A. Catalytic mechanisms of NO reduction in a CO-NO atmosphere at Co- and Cu-doped $SrTiO_3$(100) surfaces. Journal of Physical Chemistry C. 2018;**122**:449-454. DOI: 10.1021/acs.jpcc.7b09279

[13] DeZanet A, Peron G, Natile MM, Vittadini A, Glisenti A. Synthesis and development of four way catalysts starting from critical raw material free Perovskites: Influence of doping and synthesis conditions. Topics in Catalysis. 2019;**62**:237-243. DOI: 10.1007/s11244-018-1119-7

[14] Glisenti A, Pacella M, Guiotto M, Natile MM, Canu P. Largely Cu-doped $LaCo_{1-x}Cu_xO_3$ perovskites for TWC: Toward new PGM-free catalysts. Applied Catalysis B: Environmental. 2016;**180**:94-105. DOI: 10.1016/j.apcatb.2015.06.017

[15] Pacella M, Garbujo A, Fabro J, Guiotto M, Xin Q, Natile MM, et al. PGM-free $CuO/LaCoO_3$ nanocomposites: New opportunities for TWC application. Applied Catalysis B:

Environmental. 2018;**227**:446-458. DOI: 10.1016/j.apcatb.2018.01.053

[16] Grillo F, Natile MM, Glisenti A. Low temperature oxidation of carbon monoxide: The influence of water and oxygen on the reactivity of a Co_3O_4 powder surface. Applied Catalysis B: Environmental. 2004;**48**:267-274. DOI: 10.1016/j.apcatb.2003.11.003

[17] Garbujo A, Pacella M, Natile MM, Guiotto M, Fabro J, Canu P, et al. On A-doping strategy for tuning the TWC catalytic performance of perovskite based catalysts. Applied Catalysis A: General. 2017;**544**:94-107. DOI: 10.1016/j.apcata.2017.07.009

[18] Pinto D, Glisenti A. Pulsed reactivity on $LaCoO_3$-based perovskites: A comprehensive approach to go inside CO oxidation mechanism and the effect of dopants. Catalysis Science & Technology. 2019;**9**:2749-2757. DOI: 10.1039/C9CY00210C

[19] Glisenti A, Galenda A, Natile MM. $La_{0.7}Sr_{0.3}CuO_{3-\delta}$: An interesting catalyst for methanol and ethanol treatment. Catalysis Letters. 2013;**143**:254-259. DOI: 10.1007/s10562-012-0954-6

[20] Glisenti A, Galenda A, Natile MM. Steam reforming and oxidative steam reforming of methanol and ethanol: The behaviour of $LaCo_{0.7}Cu_{0.3}O_3$. Applied Catalysis A: General. 2013;**453**:102-112. DOI: 10.1016/j.apcata.2012.11.031

[21] Perin G, Fabro J, Guiotto M, Xin Q, Natile MM, Cool P, et al. Cu@$LaNiO_3$ based nanocomposites in TWC applications. Applied Catalysis B: Environmental. 2017;**209**:214-227. DOI: 10.1016/j.apcatb.2017.02.064

[22] Somorjai GA. Introduction to Surface Chemistry and Catalysis. New York: John Wiley & Sons Inc;

[23] Carollo G, Garbujo A, Xin Q, Fabro J, Cool P, Canu P, et al. CuO/$La_{0.5}Sr_{0.5}CoO_3$ nanocomposites in TWC reference: APCATB_117753. Applied Catalysis B: Environmental. in press

[24] Guiotto M, Pacella M, Perin G, Iovino A, Michelon N, Natile MM, et al. Washcoating vs. direct synthesis of $LaCoO_3$ on monoliths for environmental applications. Applied Catalysis A: General. 2015;**499**:146-157. DOI: 10.1016/j.apcata.2015.04.013

[25] Fadley CS. Angle-resolved X-ray photoelectron spectroscopy. Progress in Surface Science. 1984;**16**:275-388. DOI: 10.1016/0079-6816(84)90001-7

[26] Fadley CS. Basic Concepts of X-Ray Photoelectron Spectroscopy in Electron Spectroscopy: Theory, Techniques and Application. Vol. 2. London: Academic Press; 1978

[27] Argile C, Rhead GE. Adsorbed layer and thin film growth modes monitored by Auger electron spectroscopy. Surface Science Reports. 1989;**30**:277-356. DOI: 10.1016/0167-5729(89)90001-0

[28] Pakhare D, Spivey J. A review of dry (CO_2) reforming of methane over noble metal catalysts. Chemical Society Reviews. 2014;**43**:7813-7837. DOI: 10.1039/c3cs60395d

[29] Jang WJ, Shim J-O, Kim H-M, Yoo S-Y, Roh H-S. A review on dry reforming of methane in aspect of catalytic properties. Catalysis Today. 2019;**324**:15-26. DOI: 10.1016/j.cattod.2018.07.032

[30] Wang Y, Yao L, Wang S, Mao D, Hu C. Low-temperature catalytic CO_2 dry reforming of methane on Ni-based catalysts: A review. Fuel Processing Technology. 2018;**169**:199-206. DOI: 10.1016/j.fuproc.2017.10.007

[31] Aramouni NAK, Touma JG, Tarboush BA, Zeaiter J, Ahmad MN. Catalyst design for dry reforming of methane: Analysis review. Renewable

and Sustainable Energy Reviews. 2018; **82**:2570-2585. DOI: 10.1016/j.rser.2017.09.076

[32] Papadopoulou C, Matralis H, Verykios X. Utilization of biogas as a renewable carbon source: Dry reforming of methane. In: Catalysis for Alternative Energy Generation. New York: Springer. pp. 57-127

[33] Kumar N, Kanitkar S, Wang Z, Haynes D, Shekhawat D, Spivey JJ. Dry reforming of methane with isotopic gas mixture over Ni-based pyrochlore catalyst. International Journal of Hydrogen Energy. 2019, 2019;**44**: 4167-4176. DOI: 10.1016/j.ijhydene.2018.12.145

[34] Bachiller-Baeza B, Mateos-Pedrero C, Soria MA, Guerrero-Ruiz A, Rodemerck U, Rodríguez-Ramos I. Transient studies of low-temperature dry reforming of methane over Ni-CaO/ZrO_2-La_2O_3. Applied Catalysis B: Environmental. 2013;**129**:450-459. DOI: 10.1016/j.apcatb.2012.09.052

[35] Ye LT, Zhang MY, Huang P, Guo GC, Hong MC, Li CS, et al. Enhancing CO_2 electrolysis through synergistic control of non-stoichiometry and doping to tune cathode surface structures. Nature Communications. 2017;**8**:14785. DOI: 10.1038/ncomms14785

[36] Tien-Thao N, Alamdari H, Zahedi-Niaki MH, Kaliaguine S. Applied Catalysis A: General. 2006;**311**:204-212. DOI: 10.1016/j.apcata.2006.06.029

[37] Tien Thao N, Alamdari H, Kaliaguine S. Journal of Solid State Chemistry. 2008;**181**:2006-2019. DOI: 10.1016/j.jssc.2007.11.016

[38] Bedon A, Natile MM, Glisenti A. On the synthesis and stability of $La_{0.6}Sr_{0.4}Ga_{0.3}Fe_{0.7}O_3$. Journal of the European Ceramic Society. 2017;**37**: 1049-1058. DOI: 10.1016/j.jeurceramsoc.2016.10.017

[39] Bedon A, Rieu M, Viricelle JP, Glisenti A. Rational development of IT-SOFC electrodes based on the Nanofunctionalization of $La_{0.6}Sr_{0.4}Ga_{0.3}Fe_{0.7}O_3$ with oxides. PART 1: Cathodes by means of Iron oxide. ACS Applied Energy Materials. 2018;**1**:6840-6850. DOI: 10.1021/acsaem.8b01124

[40] Galenda A, Natile MM, Glisenti A. Oxygen permeation measurements: An alternative tool to select new intermediate temperature solid oxide fuel cell cathodes. Nanoscience and Nanotechnology Letters. 2011;**3**:723-730. DOI: 10.1166/nnl.2011.1227

[41] Natile MM, Eger G, Batocchi P, Mauvy F, Glisenti A. Strontium and copper doped $LaCoO_3$: New cathode materials for solid oxide fuel cells? International Journal of Hydrogen Energy. 2017;**42**:1724-1735. DOI: 10.1016/j.ijhydene.2016.09.127

Chapter 5

Perovskite-Based Formulations as Rival Platinum Catalysts for NO_x Removal in Diesel Exhaust Aftertreatment

Jon Ander Onrubia-Calvo, Beñat Pereda-Ayo, Unai De-La-Torre and Juan Ramón González-Velasco

Abstract

NO_x removal is still a technological challenge in diesel engines. NO_x storage and reduction (NSR), selective catalytic reduction (SCR), and combined NSR-SCR systems are the efficient approaches for diesel exhaust aftertreatment control. However, NSR and combined NSR-SCR technologies require high noble metal loadings, with low thermal stability and high cost. Recently, perovskites have gained special attention as an efficient alternative to substituting noble metals in heterogeneous catalysis. Up to date, few studies analyzed the application of perovskites in automobile catalytic converters. This chapter overviews recent research on development of novel perovskite-based catalysts as a component of single-NSR and hybrid NSR-SCR systems for NO_x removal from diesel engine exhaust gases. Results in our laboratory are compared with similar work reported in the literature by other authors. Under realistic conditions, 0.5% Pd–30% $La_{0.5}Ba_{0.5}CoO_3/Al_2O_3$ catalyst achieves NO_x-to-N_2 conversion higher than 92% when is coupled with an SCR catalyst placed downstream. The results show promise for a considerably higher thermal stability and lower cost diesel exhaust treatment system.

Keywords: perovskite, Pt-free catalyst, NO_x removal, lean-burn engine, NSR, NSR-SCR

1. Introduction

Diesel and lean-burn engines operate with high air-to-fuel (A/F) ratios in the range of 20–65 depending on the design of the engine and the type of fuel being combusted. This environment leads to a better fuel economy, with lower CO_2, CO, and HC emissions than stoichiometric gasoline engines (A/F~14.6). As a result, diesel and lean-burn engines gained popularity during the last decades, especially in Europe. However, operation under such net oxidizing environment makes three-way catalysts (TWC) not efficient enough to meet Euro VI standards regarding NO_x emissions in those engines (**Table 1**). Furthermore, the production of particulate matter, also known as soot, is still unavoidable [1, 2]. As a result, diesel and lean-burn engines require the implementation of

Type of vehicle	Euro level*	NO_x emission limit	Year of implementation	Aftertreatment system composition
Heavy-duty engines	1–3/I–III			DOC
	3/III	5[a]	2000	DOC + DPF
	4/IV	3.5[a]	2005	DOC + DPF
	5/V	2[a]	2008	DOC + DPF + $DeNO_x$
	6/VI	0.4/0.46[a]	2014	DOC + DPF + $DeNO_x$ + ASC
	1–3/I–III			DOC
Passengers vehicles	3/III	0.5[b]	2001	DOC + DPF
	4/IV	0.25[b]	2006	DOC + DPF
	5/V	0.18[b]	2011	DOC + DPF + $DeNO_x$
	6/VI	0.08[b]	2015	DOC + DPF + $DeNO_x$ + ASC

[a]*Type approval test for HDVs is conducted on an engine dynamometer, and limits defined as mass emitted per unit of mechanical work done (g kW h^{-1}).*
[b]*Type approval test for LDVs is conducted on an engine dynamometer, and limits defined as mass emitted per unit of distance driven (g km^{-1}).*
**European Union heavy-duty engine emission standards are denoted by Roman numerals, while light-duty vehicle standards are denoted by Arabic numbers.*

Table 1.
Evolution of Euro regulations for heavy-duty engines and passenger vehicles.

aftertreatment systems to control pollutant emissions, especially those related to NO_x and soot emission.

Current diesel engine exhaust treatment system can contain: (i) diesel oxidation catalyst (DOC); (ii) diesel particulate filter (DPF); (iii) NO_x reduction catalyst; and (iv) ammonia slip catalyst (ASC) [3]. NO_x storage and reduction (NSR) and selective catalytic reduction (SCR) technologies are the most promising approaches to control NO_x emission [4]. NSR system, also known as lean NO_x trap (LNT), operates cyclically under lean-rich periods with 1.5% Pt–15% BaO/Al_2O_3 as model catalyst. On the other hand, NH_3-SCR systems are based on the selectively catalyzed reduction of NO_x-to-N_2 with externally added NH_3 (produced by hydrolysis of urea) in an oxygen-rich environment. Cu or Fe/zeolite catalysts are the model NH_3-SCR formulations [5]. Current status of these technologies has some drawbacks that are limiting their extended implementation.

During the last years, a reasonable interest in linking NSR and NH_3-SCR systems is growing [6–9], because NSR systems generate NH_3 as byproduct during the short-rich period, whereas this compound is the usual selective reducing agent in the SCR technology. As a result, NO_x removal efficiency of the hybrid system increases notably with a simultaneous decrease in the NH_3 slip. Hence, the combined NSR-SCR system is considered as a potential solution to overcome main limitations of the stand-alone NSR and stand-alone SCR technologies. The hybrid NSR-SCR technology consists of two catalysts (NSR and SCR) arranged in series or in a single brick, which runs cyclically similarly to single-NSR systems. Up to now, the behavior of hybrid LNT-SCR systems has been mainly verified with the model 1.5% Pt–15% BaO/Al_2O_3 NSR catalyst. As already mentioned, the presence of platinum makes this formulation costly and limits hydrothermal stability.

Libby [10] and Voorhoeve et al. [11] proposed firstly in early 1970s a perovskite-based catalyst for automotive applications. From then, several studies were carried out related to the utilization of perovskite-based catalysts in diesel exhaust control. The perovskite formulation corresponds to oxides with ABO_3 and/or A_2BO_4 structure, where A is the larger cation located in the center edge of the structure and B is a smaller cation located in the center of the octahedron [12]. Specifically, A can be a lanthanide, alkaline, or alkaline-earth cation, and B cation can be any metallic element from 3, 4, or 5d configuration. One of the main advantages of the perovskite structure is the possibility to adopt a wide range of different compositions, changing either the A or the B cation or partially substituting each of them by other cations with same or different valences without destroying the perovskite structure. This leads to the formation of oxygen vacancies or changes in the oxidation state of A and B cations, allowing modulation of catalytic properties of the sample to better adapt to automotive applications [13].

All above in mind, the objective of this chapter is to provide a general outlook on utilization of perovskite-based formulations as stand-alone NSR catalysts as well as combined with a zeolite SCR catalyst to conform an efficient hybrid NSR-SCR system. First, a general overview of the application of perovskite-based formulations to control nitrogen oxide emissions from diesel engines is addressed. Then, the applicability of the perovskite-based formulation to single-NSR and combined NSR-SCR technologies will be emphasized. Special attention is paid to the promise and viability of this type of materials as alternative to Pt-based NSR model catalysts.

2. General overview on application of perovskite-based catalysts for NO_x emission control

Perovskite oxides exhibit a range of stoichiometry and crystal structures. In fact, they could accommodate around 90% of the metallic natural elements of the periodic table. The A and Ba cations can be partially replaced inside the structure, allowing tailoring their catalytic properties to better adapt to their application. Furthermore, physicochemical properties can be controlled by the modification of preparation method. As a result, these materials have been widely implemented in heterogeneous catalysis. Moreover, their high-hydrothermal stability enables their application in catalytic processes carried out at high temperatures [12, 13].

Many works suggest the application of perovskite oxides as alternative formulations to those based on platinum-group metals (PGMs) in automotive exhaust catalytic converters [10, 11, 14, 15]. This type of material has shown excellent activity in oxidation reaction working as diesel oxidation catalyst (DOC) [15–23]. Perovskite oxides demonstrated to be efficient for the simultaneous removal of NO_x and soot combustion in diesel engines allowing their implementation in diesel particulate-NO_x reduction filter (DPNR) [24–31]. Furthermore, NO_x decomposition in the form of nitrous oxide or nitric oxide has been proposed as a one their potential applications [32–38]. Finally, these formulations have been widely implemented for NO_x reduction in both stoichiometric gasoline engines (three-way catalyst, TWC) [24, 39–45] and diesel or lean-burn gasoline engines. Indeed, their implementation in the control of NO_x emission from diesel engines has gained special attention during the last decades, both in the selective catalytic reduction (SCR) and in the NO_x storage and reduction (NSR) systems.

2.1 Selective catalytic reduction (SCR)

SCR technology consists in the selective reduction of NO_x by different reducing agents (NH_3, H_2, or HC) in a net oxidizing environment. The NH_3-SCR alternative became as the most promising avenue for NO_x control in diesel engines. This technology was initially implemented in stationary emission sources. However, their characteristics permit to adopt it for automobile applications. SCR technology runs under steady-state operation conditions with continuous admission of NH_3 to stoichiometrically reduce NO_x in an oxygen-rich environment. A urea tank is usually required for NH_3 supply (by hydrolysis of urea) to achieve the SCR reactions. Due to the requirement of large space to house the urea tank, the implementation of this alternative is limited to heavy-duty vehicles. Another disadvantage is the need of a NH_3-slip catalyst to avoid NH_3 emission. Furthermore, the ammonia decomposition occurs above 180°C, which limits the NO_x removal efficiency at low temperatures.

It is widely accepted [46–48] that the following three main reactions occur during NO_x reduction through NH_3-SCR: (i) standard SCR ($4NH_3 + 4NO + O_2 \rightarrow 4H_2 + 6H_2O$); (ii) fast SCR ($2NH_3 + NO + NO_2 \rightarrow 2\ N_2 + 3H_2O$), and (iii) slow NO_2 SCR ($4NH_3 + 3NO_2 \rightarrow 3.5\ N_2 + 6H_2O$). The extent of these reactions depends on the NO/NO_2 ratio, which in turn is related to the oxidation capacity of the catalyst. NO_x removal efficiency is favored with NO/NO_2 ratio around 1 as promoting the fast SCR reaction [49] and occurring reaction at lower temperature. Nevertheless, side reactions such as NH_3 oxidation, NO oxidation, or N_2O formation from ammonium nitrate decomposition can also occur.

NH_3-SCR formulations have evolved from vanadia-based catalysts, first adopted in stationary sources, to the current Cu or Fe supported over new nano-pore zeolites with chabazite-type structure, such as SSZ-13 or SAPO-34. These formulations have already been implemented for NO_x emission control in heavy-duty vehicles and some recently in some passenger's cars in Europe, mainly due to high NO_x removal efficiency in a wide temperature window discovered with this small pore zeolite structure.

Figure 1 shows the NH_3-SCR behavior of a 4% Cu/SAPO-34 prepared in our laboratory by the solid state ion exchange method [50]. Experiments were carried out with a feed stream composed of 660 ppm NO, 660 ppm NH_3, 6% O_2, and Ar to balance. NO conversion increased with temperature as the NH_3-SCR reactions are promoted, reaching almost full conversion in an extended range from 200 to 350°C and decreasing afterward as the oxidation of ammonia with O_2 is favored at higher temperatures [51]. NH_3 conversion also increases with temperature, but 100% conversion was maintained above certain temperature where the NH_3-O_2 reaction prevails. Regarding selectivity toward N_2 is around 95–100% below 350°C, whereas above this temperature, it starts to decrease due to the NH_3 partial oxidation, which partially limits NO_x NH_3-SCR reactions. The excellent $DeNO_x$ activity of this formulation is attributed to the preferential presence of copper as isolated Cu^{2+} ions in the double six member rings $(d6r)^2$; however, the presence of CuO aggregates also plays an important role in the NO-to-NO_2 conversion oxidation.

Alternative compounds have been investigated for NH_3-SCR technology, such as supported metal oxides (MnO_x/Al_2O_3 and V_2O_5/activated carbon) [52–54], mixed oxides derived from hydrotalcite compounds such as Cu-Mg-Al [55], and perovskites oxides. Most of the catalytic studies related to the utilization of perovskite-type compositions in $DeNO_x$ technologies are based on La as A cation. However, few of them are related to NH_3-SCR technology, being focused most of them on a great majority on H_2-SCR and HC-SCR alternatives [3]. The NO_x removal efficiency of $LaMnO_3$, $LaMn_{0.95}V_{0.05}O_3$, and $BiMnO_3$ perovskites was analyzed [56–58]. Among them, $BiMnO_3$ perovskite achieved higher NH_3-SCR activity at lower temperatures. $LaMnO_3$/attapulguite [59] and Fe-containing perovskites [57, 60] ($LaMn_{0.95}Fe_{0.05}O_3$, $LaCo_{0.3}Fe_{0.7}O_3$, or $La_{0.8}Sr_{0.2}Fe_{1-x}Rh_xO_3$) were also analyzed. However, these formulations showed limited NO_x conversion (70–90%) or selectivity toward N_2. Taking into account the results observed in **Figure 1**, these formulations still not represent a real alternative to current Cu/chabazite NH_3-SCR catalysts.

2.2 NO_x storage and reduction (NSR)

The NSR concept, also known as lean NO_x trap (LNT), was pioneered by Toyota in the middle 1990s [61]. In this technology, the engine works predominantly feeding a fuel-lean mixture with periodical short-rich excursions. During the lean

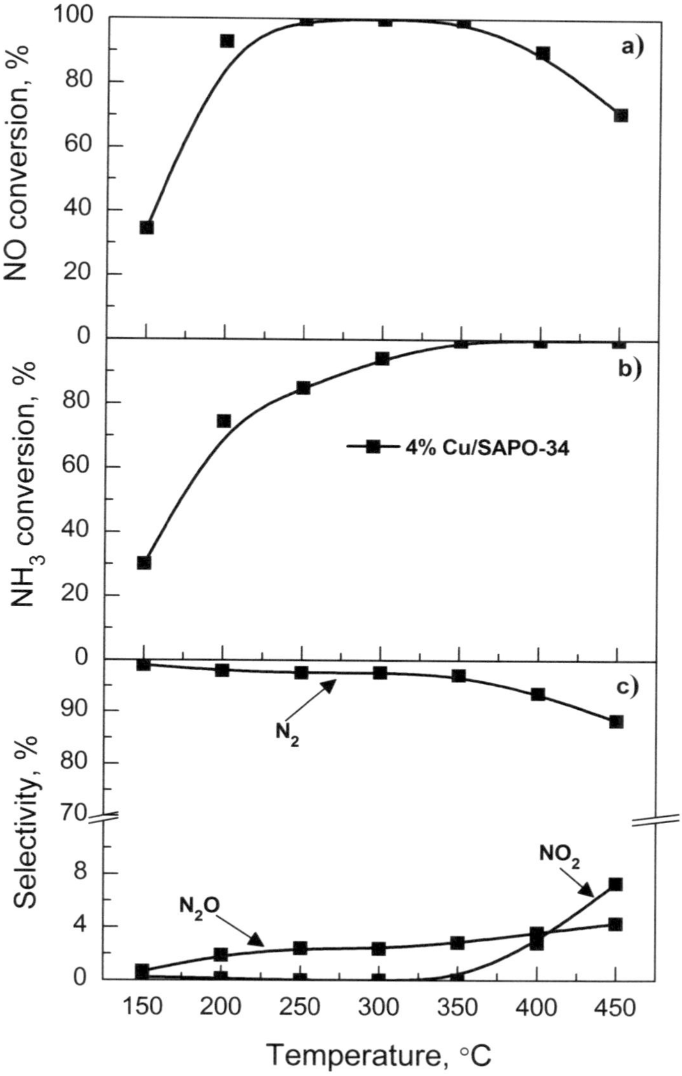

Figure 1.
Evolution of NO conversion (a), ammonia conversion (b), and selectivity toward nitrogen, nitrogen dioxide, and nitrous oxide with reaction temperature, achieved with 4% Cu/SAPO-34 catalyst. Feed: 660 ppm NO, 660 ppm NH_3, 6% O_2, Ar to balance; W/F_{Ao} = 222 (g cat.) h Mol^{-1}.

period, the NO is oxidized to NO_2 and then adsorbed over the catalyst in the form of nitrites and specially nitrates up to its saturation. Then, the stored NO_x should be released and reduced by a reductant, such as CO, H_2, or HC, during the short-rich period. The operational principle addressed the choice of NSR catalyst composition, which usually contains platinum group metals (e.g. Pt, Pd, and Rh) to activate NO oxidation and NO_x reduction and an alkaline or alkaline earth metal (e.g. K, Ba, Ca, and Sr) to promote NO_x adsorption during lean conditions. Both metals are well distributed over high-surface area materials as alumina, ceria, zirconia, or mixed oxides. A composition consisting of (1–2%) Pt/(10–15%) BaO/Al_2O_3 is widely accepted as the model NSR formulation [62–65]. **Figure 2** shows the typical NO_x storage and reduction operational principle on the NSR model catalyst [66].

LNT system shows some drawbacks derived from the operation principle and model formulation composition. On the one hand, LNT system shows NO_x leak due

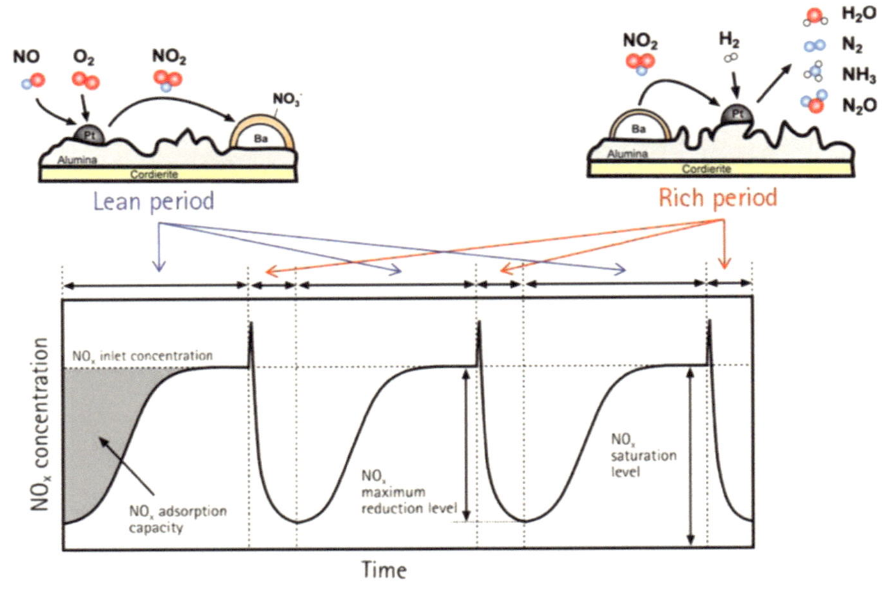

Figure 2.
NO_x storage and reduction: mechanism (upper figure), NO_x outlet concentration during three consecutive lean-rich cycles (bottom figure).

to the dynamic operation conditions under lean-rich conditions, and large amounts of N_2O and NH_3 can be also formed during rich period [67]. Furthermore, the catalyst requires high Pt loading to promote NO-to-NO_2 oxidation, which increases the cost and decreases thermal stability. Finally, the resistance to sulfur poisoning is also limited. Thus, the application of NSR technology is limited to light-duty vehicles with lean-burn engines using low-sulfur containing fuels [3].

During the last decades, modifications in composition of NSR model catalyst and new formulations such as perovskite-based materials have been explored with enhanced catalytic properties, strong deactivation resistance, and lower cost.

The application of perovskites to NSR application is mainly based on high capacity of this material to adsorb NO_x during the lean period. NO-to-NO_2 oxidation is considered a primary step for NO_x adsorption via nitrates in the model NSR catalyst, on which NO_2 adsorbs much faster than NO. With model NSR catalyst, this requires high Pt loads, which drastically increases the cost and limits the thermal stability [68, 69]. Many authors focused on development of perovskite-based formulations with high NO oxidation capacity as promising materials for use in automobile applications. In this sense, perovskite structures (ABO_3) such as $LaCoO_3$ and $LaMnO_3$ showed excellent performance on oxidation reactions [70, 71]. Choi et al. [72] reported that the catalytic oxidation activity is intimately connected to molecular and atomic interactions of oxygen with the oxide surface. Catalytic oxidation over metal oxides (M) is often rationalized in terms of a Mars-van Krevelen mechanism [73, 74], in which vacancies (□) in the oxide lattice facilitate the adsorption and dissociation of O_2.

$$2(\text{—M—}\square\text{—M—}) + O_2 \rightarrow 2(\text{—M—O—M—}) \quad (1)$$

Subsequent reaction with a reductant (R) reforms the vacancies to complete the catalytic cycle.

$$R + (\text{—M—O—M—}) \rightarrow (\text{—R—O}) + (\text{—M—}\square\text{—M—}) \quad (2)$$

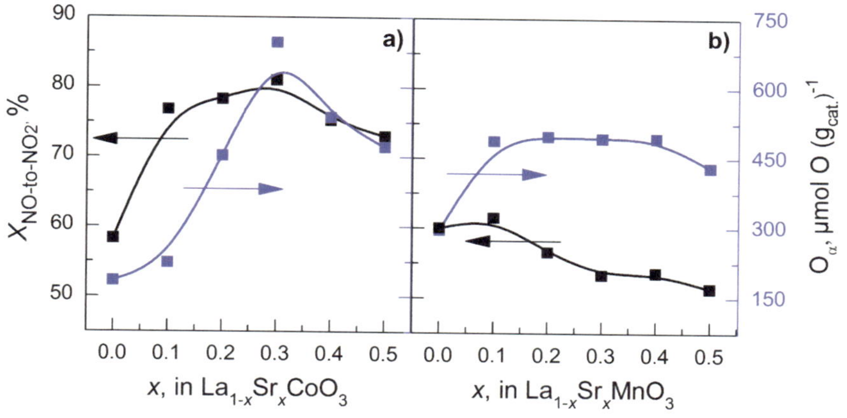

Figure 3.
Evolution of α desorbed oxygen species and NO-to-NO_2 conversion at 300°C with lanthanum substitution degree for (a) $La_{1-x}Sr_xCoO_3$ and (b) $La_{1-x}Sr_xMnO_3$ perovskites (reprinted from Ref. [16] with permission of Elsevier).

As a result, perovskite activity for oxidation reactions seems to be related to a change in the oxidation state of B cation, active oxygen mobility, and ion vacancy defect [70]. The enhancement of oxidation activity of perovskite-based catalysts is usually attributed to a promotion of oxygen vacancy density [75–79]. In this sense, lanthanum partial substitution by other cations modifies the composition and alters the physico-chemical properties of perovskite, such as crystallinity, specific surface area, average crystal size, abundance of oxygen vacancies, and oxidation state of B cation. Among different cations, Sr^{2+} seems to be the most promising cation for this approach.

Figure 3 shows the evolution of α oxygen species concentration and NO-to-NO_2 conversion at 300°C with degree of lanthanum substitution by Sr, for (a) $La_{1-x}Sr_xCoO_3$ and (**Figure 3a**) $La_{1-x}Sr_xMnO_3$ perovskites (**Figure 3b**). Note that α-oxygen was assigned to the oxygen release from vacancies located very near to or on the surface [16].

As a general trend, Sr promotes in a higher degree the formation of α oxygen species and NO-to-NO_2 oxidation capacity for Co-based perovskites than for Mn ones. The evolution of NO-to-NO_2 conversion with lanthanum substitution degree confirms that the amount of oxygen vacancies is the key factor for this enhancement. As a result, Co-based perovskites show higher NO oxidation capacity, even above than Pt-based catalyst does [16]. These results confirm that perovskites can be considered as an excellent alternative for promotion NO oxidation reactions in automotive catalysis.

Nonetheless, $La_{0.7}Sr_{0.3}CoO_3$ tends to agglomerate under high temperatures required during the calcination step (**Figure 4**). Thus, low specific surface areas (around 20 $m^2\ g^{-1}$) and an insufficient number of NO_x storage sites [80, 81] arise as main drawbacks of bulk perovskites. Two approaches have been proposed to overcome this limitation: synthesizing mesostructured perovskites via nanocasting and/or distribution of perovskite over high-surface area materials [12]. Mesoporous supports were tried in the past. In this sense, overlaying $ZrTiO_4$ with $LaCoO_3$ perovskite was found to reduce sintering of perovskite, which improves NO_x storage capacity [82]. More recently, You et al. [83, 84] found that ceria-supported and $Ce_{0.75}Zr_{0.25}O_2$-supported $LaCoO_3$ perovskite achieved high NO_x storage and reduction capacity even with low-specific surface area (below 50 $m^2\ g^{-1}$). We analyzed the effect of incorporating increasing loadings of $La_{0.7}Sr_{0.3}CoO_3$ perovskite over a conventional alumina support [85], which inhibited crystal growth of

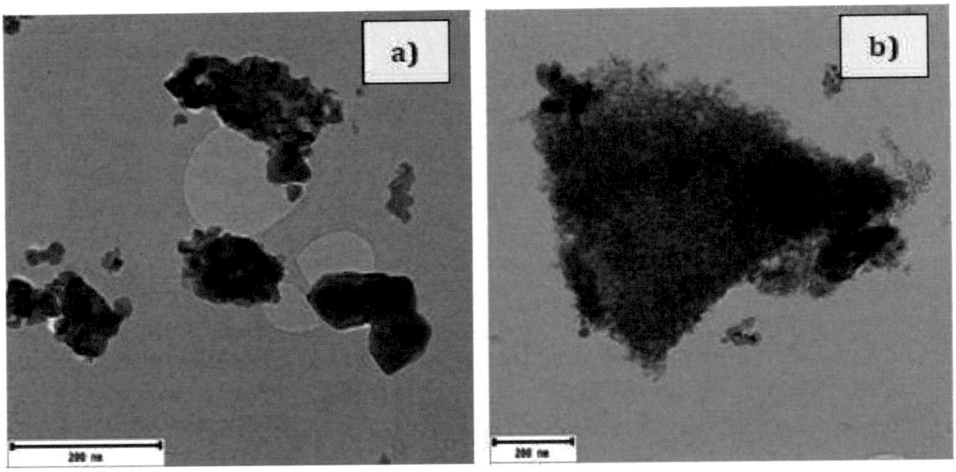

Figure 4.
TEM images of: (a) $La_{0.7}Sr_{0.3}CoO_3$ and (b) 30% $La_{0.7}Sr_{0.3}CoO_3/Al_2O_3$ samples (reprinted from Ref. [85] with permission of Elsevier).

bulk perovskites (**Figure 4**). Hence, diffusion of intermediate compounds from oxidation to adsorption sites was facilitated. Among all prepared catalysts, 30% $La_{0.7}Sr_{0.3}CoO_3/Al_2O_3$ sample achieved the most efficient use of perovskite phase due to the best balance between well-developed perovskite phase and NO oxidation and NO adsorption site distribution such as oxygen vacancies, structural La and Sr at the surface, and segregated $SrCO_3$ [86, 87].

However, NO_x reduction capacity of supported formulations is still limited (**Figure 5**). The incorporation of Pd is analyzed as a promising avenue to improve the NO_x reduction capacity of the 30% $La_{0.7}Sr_{0.3}CoO_3/Al_2O_3$ catalyst. Two approaches can be used for the incorporation of palladium in the perovskite-based formulations via impregnation [88–90] and/or by doping the perovskite structure [86, 91, 92]. The former promotes palladium accessibility; meanwhile, the latter seems to prevent the metal from agglomeration during reduction steps [93, 94]. However, contradictory conclusions have been extracted about which of them is the optimum alternative [95, 96]. In a recent study, Zhao et al. [97] compared both Pd incorporation methods for $La_{0.7}Sr_{0.3}CoO_3$ perovskite. In their study, NO_x adsorption during lean conditions and NO_x reduction to N_2 during rich period is significantly promoted after the incorporation of Pd, especially by impregnation method. The enhancement of the catalytic performance is related to a higher NO_x adsorption site regeneration and to a promotion of NO_x reduction rate by the palladium incorporation, respectively. In our previous work, we prepared several catalysts with increasing palladium contents (0.75, 1.5, and 3.0%) incorporated doping perovskite structure or by wetness impregnation over alumina-supported perovskite. They concluded that the 1.5% Pd–30% $La_{0.7}Sr_{0.3}CoO_3/Al_2O_3$ sample shows the best balance between NO_x removal efficiency and minimum palladium content. The NO_x removal efficiency and nitrogen production are as high as 86.2 and 69.5%, respectively (**Figure 5**). DeNO_x activity of this formulation is similar or even higher than that achieved with the reference catalyst (1.5% Pt–15% BaO/Al_2O_3). Thus, the developed formulation revealed as a promising alternative to the NSR model catalyst for NO_x removal in the automotive application.

It is worth noting that proposed alternative showed a high NO_2 outlet concentration under oxidizing conditions [66]. This suggests that on these materials more amount of NO_2 is formed than the catalyst can adsorb during the lean period.

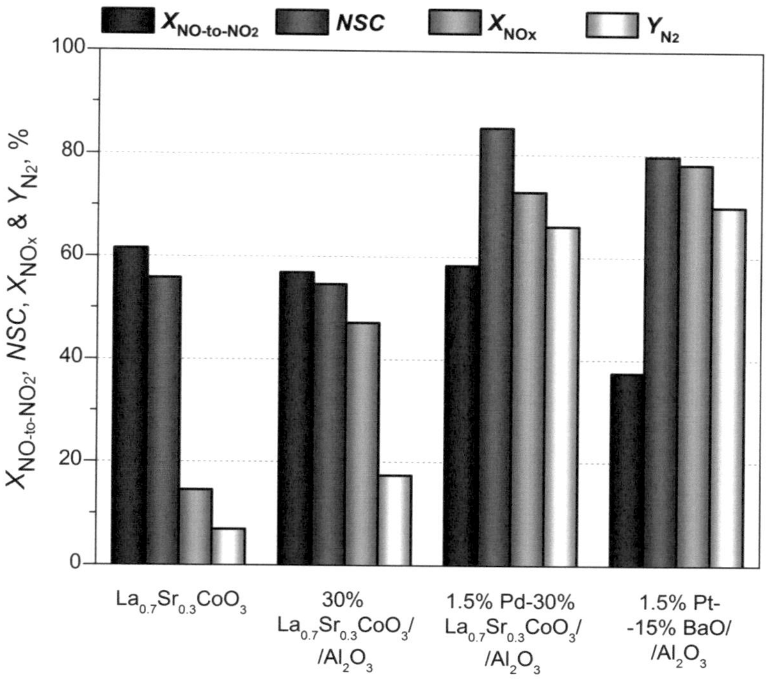

Figure 5.
NO-to-NO_2 conversion ($X_{NO\text{-}to\text{-}NO2}$, first column), NO_x storage capacity (NSC, second column), global NO_x conversion (X_{NOx}, third column), and nitrogen production (Y_{N2}, fourth column) at 400°C for perovskite-based formulations and NSR model catalyst. Feed: 500 ppm NO, 6% O_2 (lean)/3% H_2 (rich), Ar to balance; W/F_{Ao} = 200 (g cat.) h Mol^{-1}.

Furthermore, the amount of NO_x released during the rich period denotes low stability of adsorbed species, which induces fast NO_x release when the reductant is injected. Thus, NO_x reduction, and as a consequence N_2 production, could be further promoted. The increase of the concentration and strength of NO_x adsorption sites by controlling an adequate balance between NO oxidation capacity and NO_x adsorption site concentration and strength at the surface [80, 81] could be an alternative to overcome observed limitations. Two alternatives have been explored: (i) incorporation of additional NO_x adsorption sites [82–84, 98, 99] and (ii) modification of perovskite composition to alter the nature and surface concentration of NO_x adsorption sites [86, 100, 101]. The reported results show that the increase in NO_x adsorption site concentration promotes NO_x storage capacity confirming that the gas/solid equilibrium between NO_2 and the available NO_x adsorption sites is a key factor to maximize *NSC*; meanwhile, the higher strength of adsorber species favors NO_x reduction efficiencies during short-rich period. Thus, both alternatives improved the catalytic behavior of the corresponding perovskite-based formulation. In our case, the selected approach was the modification of the perovskite composition by Ba doping instead of Sr doping. The developed formulation 0.5% Pd–30% $La_{0.5}Ba_{0.5}CoO_3/Al_2O_3$ adsorbs NO_x in the form of nitrites/nitrates over surface basic sites, such as La or Ba-terminated perovskite surface, segregated $BaCO_3$, or alumina support during lean conditions. Then, adsorbed NO_x is released and reduced over Pd and in lower extent perovskite sites to form nitrogen containing products, such as N_2O, NH_3, or N_2. Furthermore, a slower reaction of the NH_3 formed with the stored nitrates leading to the selective formation of N_2 also takes place [66].

A critical aspect of NSR model catalyst (1.5% Pt–15% BaO/Al_2O_3) is the low sulfur resistance due to the formation of stable barium sulfate, which limits NO_x

adsorption during lean conditions. Hodjati et al. [102] analyzed NO_x storage performance of ABO_3 perovskite-type catalysts (with A = Ca, Sr, or Ba; and B = Sn, Zr, or Ti). Regarding A-site cations, the NO_x storage capacity (*NSC*) followed the order Ba > Sr > Ca, whereas in the case of B cation, the order was Sn > Zr > Ti. Nevertheless, the $BaSnO_3$ formulation exhibited limited sulfur resistance. In this sense, a $BaFeO_3$ catalyst developed latter by Xian et al. [87, 103, 104] showed a lower decrease of *NSC* after sulfating (about 11–12%). The incorporation of Ti improves sulfur resistance in a higher extent; activity decreased only 5.1% after SO_2-pretreatment of a $BaFe_{1-x}Ti_xO_3$ catalyst (x = 0.1 or 0.2).

In the case of La-based perovskites, $LaCo_{0.92}Pt_{0.08}O_3$ maintained a high NO_x removal efficiency after regeneration of a pre-sulfated sample [105]. $La_{0.7}Sr_{0.3}Co_{0.8}Fe_{0.2}O_3$ perovskite suffers from a drop in NO_x removal efficiency after SO_2-pretreatment of 6.4% [106]. Wang et al. [107] and Wen et al. [99] compared the sulfur and hydrothermal aging resistance of $LaCo_{0,92}Pt_{0,08}O_3$ and 0.3% Pt/(Al_2O_3 + $LaCoO_3$) catalysts, with respect to those shown by 1% Pt–16% Ba/Al_2O_3 model formulation. These alternatives achieved NO_x-to-N_2 reduction, sulfur resistance, regeneration, and durability similar or even higher than the model catalyst.

In summary, perovskite-based formulations achieve notable NO oxidation and NO_x adsorption during oxidizing conditions. Furthermore, despite the fact that only a few works analyzed NO_x removal during the short reducing period, our results summarized in this chapter remark the potential of perovskite-based materials for application in NO_x storage and reduction (NSR) technology for NO_x control in diesel and lean-burn engines. In fact, the excellent sulfur tolerance and hydrothermal resistance reported in previous work make these formulations even more promising alternative to Pt-based NSR model catalyst.

2.3 NSR-SCR combined system

As previously observed, stand-alone SCR and NSR systems have some disadvantages that hinder their extended application in both light-duty and heavy-duty vehicles. In the case of the NSR system the high cost, poor thermal stability due to the use of precious metals and nondesired byproduct generation is the main disadvantages, whereas SCR systems require an urea system to provide NH_3 and additional device to avoid ammonia slip under transient vehicle operation. The coupling of NSR and SCR catalysts has been rapidly accepted as a potential solution, since its discovery by the Ford Motor company [9, 108]. Different catalytic formulations, system architectures, and operation control have been explored [7, 8, 50, 109, 110]. The systems based on model NSR formulation and Cu/chabazite-type zeolites emerge as the most efficient combination [111]. This hybrid technology has been demonstrated more efficient by maximizing NO_x-to-N_2 reduction and minimizing NH_3 slip with respect to the alone-NSR catalyst. Nevertheless, the most studied NSR formulation used in the combined NSR-SCR system has usually been the conventional Pt-based model catalyst (1.5% Pt-15% BaO/Al_2O_3), which transfers its high cost and limited hydrothermal stability to the hybrid configuration. Based on the results demonstrated by perovskite-based formulations in the single-NSR technology, their application in combined NSR-SCR systems is considered as an evolution of the current NSR-SCR architecture.

Figure 6 shows the NO_x ($NO+NO_2$), N_2O, and NH_3 concentration profiles determined by FTIR for the single-NSR and double NSR-SCR configurations at 300°C. The N_2 signal determined by mass spectroscopy is also included. NSR and SCR formulations correspond to 0.5% Pd–30% $La_{0.5}Ba_{0.5}CoO_3$/Al_2O_3 and 4% Cu/SAPO-34 catalysts, respectively.

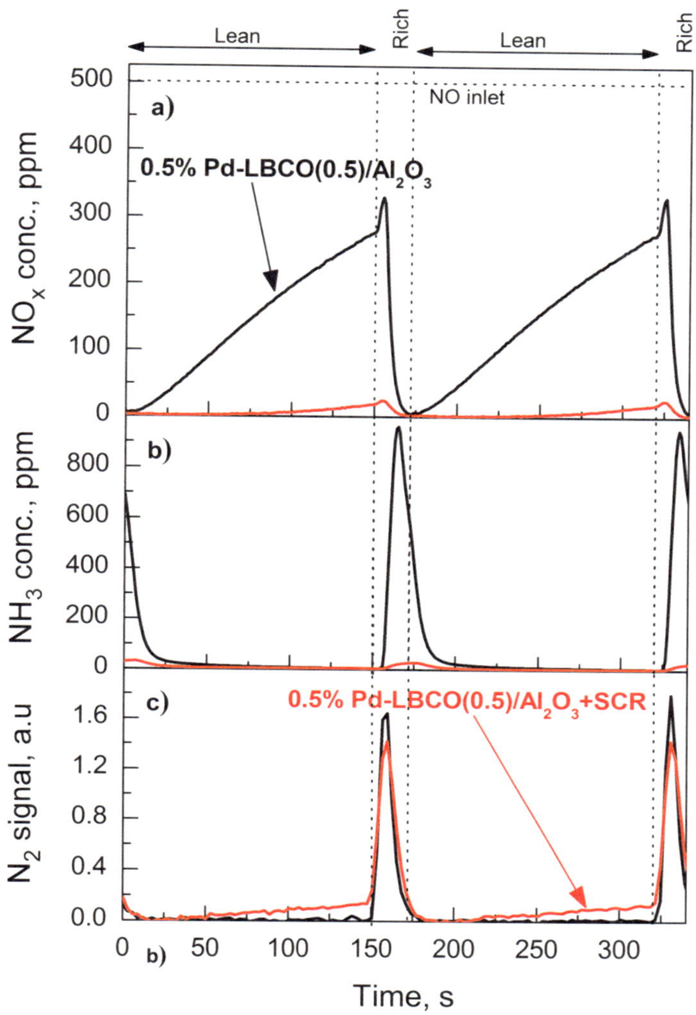

Figure 6.
NO_x ($NO + NO_2$) and NH_3 outlet concentrations, and MS signal of N_2 for the single NSR and NSR-SCR configurations at 300°C. Feed: 500 ppm NO, 6% O_2/3% H_2, and Ar to balance; W/FA_0 = 200 (g h Mol^{-1}).

The single-NSR system (**Figure 6a**) shows the typical NO_x outlet concentration profile [112]. At the beginning of the lean period, practically all NO_x fed are stored, and therefore, concentration of NO_x at the reactor outlet is almost null. Then, as the length of the lean period increases, NO_x adsorption sites become progressively saturated and NO_x outlet concentration increases. In the subsequent rich period, H_2 injected releases the stored NO_x release and reduces it to a mixture of N_2, NH_3 and N_2O [113]. During this period, the NH_3 outlet concentration peaks to almost 1000 ppm (**Figure 6b**). NO_x outlet concentration (**Figure 6a**) decreases drastically when the system operates under the NSR-SCR double configuration. As can be observed in **Figure 6b**, most NH_3 formed during regeneration of the NSR catalyst is adsorbed on acidic sites of SAPO-34, since concentration of NH_3 detected at the outlet of the combined NSR-SCR system is almost zero. Then, in the subsequent lean period, NO_x slipping the NSR catalyst reacts with NH_3 previously adsorbed on the SCR, leading to further NO_x reduction [114].

The evolution of N_2 signal also confirms the existence of SCR reaction over the Cu/SAPO-34 catalyst (**Figure 6c**). When the operation is performed with alone-NSR catalyst, formation of N_2 was only detected during the rich period, the signal being constant and negligible throughout the storage period. By contrast, when the reaction was carried out with the combined NSR-SCR system, N_2 formation was also detected during the storage period. At the beginning of this period, practically all NO_x fed are trapped in the NSR catalyst, and therefore, there is no available NO_x at the outlet gas stream to carry out the SCR reaction downstream. As a result, the N_2 signals at the exit of the NSR catalyst and at the outlet of the NSR-SCR are coincident. As the storage period proceeds, the NSR catalyst becomes saturated, and the gradual increase of NO_x concentration at intermediate stream that feeds the SCR catalyst activates its reduction with the NH_3 previously stored over the Cu/SAPO-34 catalyst to form the effluent N_2 [66].

Thus, the beneficial effect of placing an SCR catalyst downstream of the NSR is demonstrated. Hence, the catalytic behavior of different perovskite-based formulations was compared in a wider range of temperature using a H_2 concentration of 3% during the rich period. **Figure** 7 quantifies the evolution of NO conversion and N_2, NO_2, and NH_3 productions at 200, 300, and 400°C, for single-NSR and combined NSR-SCR systems. The NSR formulation was varied between 1.5% Pt–15% BaO/ Al_2O_3 (model catalyst) and 0.5% Pd–30% $La_{0.5}Ba_{0.5}CoO_3/Al_2O_3$ (perovskite-based catalyst here formulated).

As a general trend, NO conversion improves with the NSR-SCR configuration irrespective of the NSR catalyst composition used. The implementation of the SCR catalyst in the hybrid configuration also improves production of N_2, by consumption of NH_3 and NO_2 produced in the NSR catalyst. The improvement of NO_x removal efficiency up to 300°C is due to higher NH_3 production when the SCR catalyst is highly efficiency (200–300°C). In good agreement with this, NH_3 production at the outlet decreases significantly for the combined NSR-SCR systems [66]. However, above 300°C, the NH_3 production in the NSR catalyst decreases significantly due to the reaction of the NH_3 with the stored nitrate downstream ($3Ba(NO_3)_2 + 10NH_3 \rightarrow 3BaO + 8\ N_2 + 15H_2O$) of the NSR catalyst [65, 115]. Furthermore, NH_3 can be partially oxidized [113, 116], and thus, NH_3 generated is insufficient to reduce NO_x slipping the upstream NSR system. This explains the

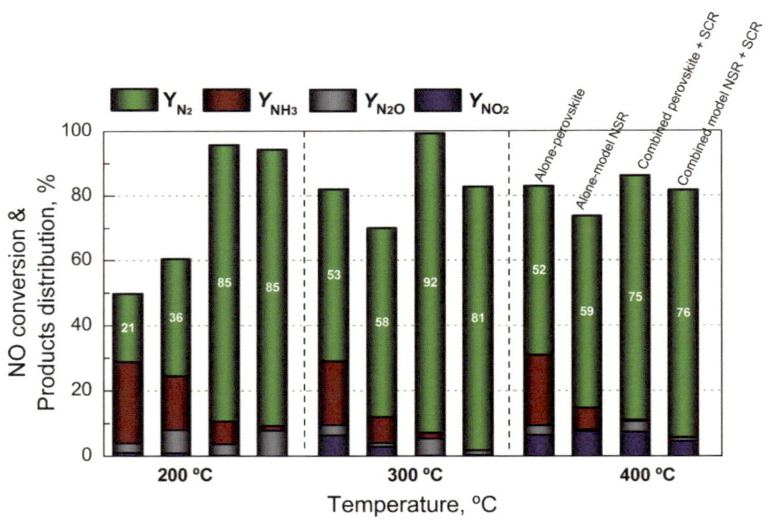

Figure 7.
Global conversion of NO (X_{NO}) and product distribution at 200, 300, and 400°C for single-NSR and combined NSR-SCR systems, based on 0.5% Pd–30% $La_{0.5}Ba_{0.5}CoO_3/Al_2O_3$ (perovs.) and 1.5% Pt–15% BaO/Al_2O_3 (model) as NSR catalysts.

moderate improvement of NO_x-to-N_2 reduction for the NSR-SCR double configuration with respect to the single-NSR system at high temperature.

Both configurations show high NO conversion and nitrogen production in the whole temperature range. Specifically, perovskite-based system shows a maximum NO conversion of 99% and a nitrogen production of 92% at 300°C. Indeed, perovskite-based combined NSR-SCR system shows similar or even higher NO_x-to-N_2 reduction efficiencies to that shown by the double configuration when the 1.5% Pt–15% BaO/Al_2O_3 model NSR catalyst is used.

In summary, the positive impact of placing an SCR catalyst downstream the NSR catalyst is verified on NO_x removal efficiency, with notable increase in N_2 production. In this sense, the mixed NSR-SCR system based on the 0.5% Pd–30% $La_{0.5}Ba_{0.5}CoO_3/Al_2O_3$ catalyst emerges as a promising alternative, only emitting 7% NH_3 at 200°C (slip of NH_3), which disappears at higher temperatures achieving NH_3 total elimination. This can be considered as a promising starting point for the implementation of these types of oxides in coupled NSR-SCR configurations.

2.4 Outlook and concluding remarks

NO_x emission removal from lean-burn and diesel engine exhaust gases remains as a technological challenge. To overcome this environmental pollution issue, two main alternatives have been explored during the last decades: NO_x storage and reduction (NSR) and selective catalytic reduction with NH_3 (NH_3-SCR). These alternatives show some limitations that limit their extensive application. In the case of NSR system, some NO can slip without being totally converted, and also, NH_3 generated during the rich period can slip as byproduct in the effluent. Furthermore, the catalyst requires high Pt loadings, which limit the cost and thermal stability. On the other hand, NH_3-SCR system requires the urea feeding system, which increases the cost and requires an extra volume of the system. Moreover, the latter shows lower NO conversion at low temperatures and allows NH_3 slip. As a result, combined NSR-SCR configurations have been explored as an evolution of previous stand-alone technologies. In fact, this hybrid alternative increases the temperature operational window, promotes NO conversion, and avoids the need of urea feeding system. However, up to now, only a conventional Pt-based NSR formulation has been explored in coupled NSR-SCR configurations.

In recent years, efforts have been focused on designing a new generation of NSR catalysts with improved oxidation, adsorption, and reduction capacities. Furthermore, these new materials should be low cost and achieve long hydrothermal stability and high sulfur resistance. Perovskites have gained attention during the recent years as a potential solution. La-based formulations (i.e., $LaCoO_3$ and $LaMnO_3$) have shown excellent NO oxidation conversion, a primary step in the NO_x adsorption during lean conditions. In fact, Sr doping further promotes the NO oxidation activity of these formulations, which is closely related to the generation of oxygen vacancies favoring oxygen mobility. However, NO_x storage and reduction efficiencies are limited for bulk perovskites due to a low exposed surface area derived from the drastic calcination protocols required during the synthesis process. Supporting perovskite over high surface area materials, e.g., alumina (30% $La_{0.7}Sr_{0.3}CoO_3/Al_2O_3$), is demonstrated to overcome this limitation. Nonetheless, NO_x reduction at low and intermediate temperatures is still limited. The incorporation of low Pd contents over supported perovskite by wetness impregnation emerges as an efficient solution. In fact, 1.5% Pd–30% $La_{0.7}Sr_{0.3}CoO_3/Al_2O_3$ shows similar or even higher NO_x removal efficiencies than the conventional NSR model catalyst (1.5% Pt–15% BaO/Al_2O_3). The activity enhancement showed by perovskite-based formulations motivates their implementation in combined

NSR-SCR systems, which as an alternative to further improve the NO_x removal efficiency of the stand-alone NSR and stand-alone SCR systems. The preliminary results are very promising since NO_x-to-N_2 reduction above 90% has been achieved with significant lower noble metal content than platinum in the model catalyst.

Improving the exhaust aftertreatment systems is considered as a critical point in the current vehicle development. In upcoming years, research should be focus on better understanding the mechanism over perovskite-based formulation, especially during regeneration period. Moreover, the NO_x trapping efficiency and NO_x reduction of the adsorbed NO_x can be further promoted. To the best of the authors' knowledge, no studies have been published related to the application of this type of materials to the combined NSR-SCR system. Thus, the room improvement is huge for this application, such as exploring different catalyst architectures (i.e., segmented zones or dual layer monoliths), optimizing precious metal loading, and dispersion. The construction of detailed kinetic model and modeling a full-scale operation will allow to develop a suitable aftertreatment system for automobile application.

Acknowledgements

Support from the Spanish Ministry of Economy and Competiveness (Project CTQ2015-67597-C2-1-R), the Basque Government (IT657-13 and IT1297-19), and the University of the Basque Country acknowledged. One of the authors (JAOC) was supported by a PhD research fellowship provided by the Basque Government (PRE_2014_1_396).

Author details

Jon Ander Onrubia-Calvo, Beñat Pereda-Ayo, Unai De-La-Torre and Juan Ramón González-Velasco*
Chemical Technologies for Environmental Sustainability TQSA, Department of Chemical Engineering, Faculty of Science and Technology, University of the Basque Country UPV/EHU, Leioa, Bizkaia, Spain

*Address all correspondence to: juanra.gonzalezvelasco@ehu.es

References

[1] González-Velasco JR, Pereda-Ayo B, De-La-Torre U, Urrutxua M, López-Fonseca R. Coupled NSR-SCR systems for NO_x removal in light-duty vehicles. ChemCatChem. 2018;**10**:2928. DOI: 10.1002/cctc.201800392

[2] Granger P, Parvulescu VI. Catalytic NO_x abatement systems for mobile sources: From three-way to lean burn after-treatment technologies. Chemical Reviews. 2011;**111**(5):3155-3207. DOI: 10.1021/cr100168g

[3] Jabłońska M, Palkovits R. Perovskite-based catalysts for the control of nitrogen oxide emissions from diesel engines. Catalysis Science & Technology. 2019;**9**(9):2057-2077. DOI: 10.1039/C8CY02458H

[4] Granger P. Challenges and breakthrough in post-combustion catalysis: How to match future stringent regulations? Catalysis Science & Technology. 2017;7:5195-5211. DOI: 10.1039/C7CY00983F

[5] De-La-Torre U, Pereda-Ayo B, Onrubia JA, González-Velasco JR. Effect of the presence of ceria in the NSR catalyst on the hydrothermal resistance and global $DeNO_x$ performance of coupled LNT-SCR systems. Topics in Catalysis. 2018;**61**(18):1993-2006. DOI: 10.1007/s11244-018-1016-0

[6] Gandhi HS, Cavatalo JV, Hammerle RH, Chen Y. Catalyst system for NOx and NH3 emission. US Patent 7332135 (19-02-2008)

[7] Gandhi HS, Cavatalo JV, Hammerle RH, Chen Y. Catalyst system for NOx and NH3 emission. US Patent Appl. 2004/0076565 A1 (22-04-2004)

[8] Guenther J, Konrad B, Krutzsch B, Nolte A, Voigtlaender D, Weibel M, et al. Exhaust gas purification process and apparatus with internal generation of ammonia for reducing nitrogen oxide. US Patent 6338244 B1 (15-01-2002)

[9] Kinugasa Y, Igarashi K, Itou T, Suzuki N, Yaegashi T, Takeuchi K. Method adsprobed and device for purifying exhaust gas from engine. US Patent 6047542 (11-04-2000)

[10] Libby WF. Promising catalyst for auto exhaust. Science. 1971;**171**(3970):499. DOI: 10.1126/science.171.3970.499

[11] Voorhoeve RJH, Remeika JP, Freeland PE, Matthias BT. Rare-earth oxides of manganese and cobalt rival platinum for the treatment of carbon monoxide in auto exhaust. Science. 1972;**177**(4046):353. DOI: 10.1126/science.177.4046.353

[12] Zhu J, Li H, Zhong L, Xiao P, Xu X, Yang X, et al. Perovskite oxides: Preparation, characterizations, and applications in heterogeneous catalysis. ACS Catalysis. 2014;**4**(9):2917-2940. DOI: 10.1021/cs500606g

[13] Royer S, Duprez D, Can F, Courtois X, Batiot-Dupeyrat C, Laassiri S, et al. Perovskites as substitutes of noble metals for heterogeneous catalysis: Dream or reality. Chemical Reviews. 2014;**114**(20):10292-10368. DOI: 10.1021/cr500032a

[14] Gallagher PK, Johnson DW, Schrey F. Studies of some supported perovskite oxidation catalysts. Materials Research Bulletin. 1974;**9**(10):1345-1352. DOI: 10.1016/0025-5408(74)90057-9

[15] Kim CH, Qi G, Dahlberg K, Li W. Strontium-doped perovskites rival platinum catalysts for treating NO_x in simulated diesel exhaust. Science. 2010;**327**(5973):1624-1627. DOI: 10.1126/science.1184087

[16] Onrubia JA, Pereda-Ayo B, De-La-Torre U, González-Velasco JR. Key factors in Sr-doped $LaBO_3$ (B=Co or Mn) perovskites for NO oxidation in efficient diesel exhaust purification. Applied Catalysis B: Environmental. 2017;**213**:198-210. DOI: 10.1016/j.apcatb.2017.04.068

[17] Zhou C, Feng Z, Zhang Y, Hu L, Chen R, Shan B, et al. Enhanced catalytic activity for NO oxidation over Ba doped $LaCoO_3$ catalyst. RSC Advances. 2015;**5**(36):28054-28059. DOI: 10.1039/C5RA02344K

[18] Wen Y, Zhang C, He H, Yu Y, Teraoka Y. Catalytic oxidation of nitrogen monoxide over $La_{1-x}Ce_xCoO_3$ perovskites. Catalysis Today. 2007;**126**(3):400-405. DOI: 10.1016/j.cattod.2007.06.032

[19] Wang J, Su Y, Wang X, Chen J, Zhao Z, Shen M. The effect of partial substitution of Co in $LaMnO_3$ synthesized by sol–gel methods for NO oxidation. Catalysis Communications. 2012;**25**:106-109. DOI: 10.1016/j.catcom.2012.04.001

[20] Zhou C, Liu X, Wu C, Wen Y, Xue Y, Chen R, et al. NO oxidation catalysis on copper doped hexagonal phase $LaCoO_3$: A combined experimental and theoretical study. Physical Chemistry Chemical Physics. 2014;**16**(11):5106-5112. DOI: 10.1039/C3CP54963A

[21] Zhong S, Sun Y, Xin H, Yang C, Chen L, Li X. NO oxidation over Ni-Co perovskite catalysts. Chemical Engineering Journal. 2015;**275**:351-356. DOI: 10.1016/j.cej.2015.04.046

[22] Hong Z, Wang Z, Li X. Catalytic oxidation of nitric oxide (NO) over different catalysts: An overview. Catalysis Science & Technology. 2017;7(16):3440-3452. DOI: 10.1039/C7CY00760D

[23] Zhu J, Thomas A. ChemInform abstract: Perovskite-type mixed oxides as catalytic material for NO removal. Applied Catalysis B: Environmental. 2010;**41**:225-233. DOI: 10.1016/j.apcatb.2009.08.008

[24] Hong S, Lee G. Simultaneous removal of NO and carbon particulates over lanthanoid perovskite-type catalysts. Catalysis Today. 2000;**63**(2):397-404. DOI: 10.1016/S0920-5861(00)00484-3

[25] Liu J, Xu J, Zhao Z, Duan A, Jiang G, Jing Y. A novel four-way combining catalysts for simultaneous removal of exhaust pollutants from diesel engine. Journal of Environmental Sciences. 2010;**22**(7):1104-1109. DOI: 10.1016/S1001-0742(09)60224-2

[26] Li Z, Meng M, Li Q, Xie Y, Hu T, Zhang J. Fe-substituted nanometric $La_{0.9}K_{0.1}Co_{1-x}Fe_xO_{3-\delta}$ perovskite catalysts used for soot combustion, NO_x storage and simultaneous catalytic removal of soot and NO_x. Chemical Engineering Journal. 2010;**164**(1):98-105. DOI: 10.1016/j.cej.2010.08.036

[27] Mescia D, Caroca JC, Russo N, Labhsetwar N, Fino D, Saracco G, et al. Towards a single brick solution for the abatement of NO_x and soot from diesel engine exhausts. Catalysis Today. 2008;**137**(2):300-305. DOI: 10.1016/j.cattod.2007.11.010

[28] Teraoka Y, Kanada K, Kagawa S. Synthesis of $La_{1-x}K_xMnO$ perovskite-type oxides and their catalytic property for simultaneous removal of NOx and diesel soot particulates. Applied Catalysis B: Environmental. 2001;**34**(1):73-78. DOI: 10.1016/S0926-3373(01)00202-8

[29] Yao W, Wang R, Yang X. $LaCo_{1-x}Pd_xO_3$ perovskite-type oxides: Synthesis, characterization and simultaneous removal of NO_x and diesel soot. Catalysis Letters. 2009;**130**(3):613-621. DOI: 10.1007/s10562-009-9905-2

[30] Wang K, Qian L, Zhang L, Liu H, Yan Z. Simultaneous removal of NO_x and soot particulates over $La_{0.7}Ag_{0.3}MnO_3$ perovskite oxide catalysts. Catalysis Today. 2010;**158**(3):423-426. DOI: 10.1016/j.cattod.2010.06.001

[31] Dhal GC, Dey S, Mohan D, Prasad R. Study of Fe, Co, and Mn-based perovskite-type catalysts for the simultaneous control of soot and NO_X from diesel engine exhaust. Materials Discovery. 2017;**10**:37-42. DOI: 10.1016/j.md.2018.04.002

[32] Sun K, Xia H, Hensen E, van Santen R, Li C. Chemistry of N_2O decomposition on active sites with different nature: Effect of high-temperature treatment of Fe/ZSM-5. Journal of Catalysis. 2006;**238**(1):186-195. DOI: 10.1016/j.jcat.2005.12.013

[33] Ivanov DV, Sadovskaya EM, Pinaeva LG, Isupova LA. Influence of oxygen mobility on catalytic activity of La–Sr–Mn–O composites in the reaction of high temperature N_2O decomposition. Journal of Catalysis. 2009;**267**(1):5-13. DOI: 10.1016/j.jcat.2009.07.005

[34] Ivanov DV, Pinaeva LG, Isupova LA, Sadovskaya EM, Prosvirin IP, Gerasimov EY, et al. Effect of surface decoration with $LaSrFeO_4$ on oxygen mobility and catalytic activity of $La_{0.4}Sr_{0.6}FeO_{3-\delta}$ in high-temperature N_2O decomposition, methane combustion and ammonia oxidation. Applied Catalysis A: General. 2013;**457**:42-51. DOI: 10.1016/j.apcata.2013.03.007

[35] Wu Y, Dujardin C, Lancelot C, Dacquin JP, Parvulescu VI, Cabié M, et al. Catalytic abatement of NO and N_2O from nitric acid plants: A novel approach using noble metal-modified perovskites. Journal of Catalysis. 2015;**328**:236-247. DOI: 10.1016/j.jcat.2015.02.001

[36] Gunasekaran N, Rajadurai S, Carberry JJ. Catalytic decomposition of nitrous oxide over perovskite type solid oxide solutions and supported noble metal catalysts. Catalysis Letters. 1995;**35**(3):373-382. DOI: 10.1007/BF00807194

[37] Pan KL, Yu SJ, Yan SY, Chang MB. Direct N_2O decomposition over La_2NiO_4-based perovskite-type oxides. Journal of the Air & Waste Management Association. 2014;**64**(11):1260-1269. DOI: 10.1080/10962247.2014.941513

[38] Alini S, Basile F, Blasioli S, Rinaldi C, Vaccari A. Development of new catalysts for N_2O-decomposition from adipic acid plant. Applied Catalysis B: Environmental. 2007;**70**(1):323-329. DOI: 10.1016/j.apcatb.2005.12.031

[39] Belessi VC, Trikalitis PN, Ladavos AK, Bakas TV, Pomonis PJ. Structure and catalytic activity of $La_{1-x}FeO_3$ system (x=0.00, 0.05, 0.10, 0.15, 0.20, 0.25, 0.35) for the NO+CO reaction. Applied Catalysis A: General. 1999;**177**. DOI: 10.1016/S0926-860X(98)00256-7

[40] Belessi VC, Bakas TV, Costa CN, Efstathiou AM, Pomonis PJ. Synergistic effects of crystal phases and mixed valences in La–Sr–Ce–Fe–O mixed oxidic/perovskitic solids on their catalytic activity for the NO+CO reaction. Applied Catalysis B: Environmental. 2000;**28**(1):13-28. DOI: 10.1016/S0926-3373(00)00159-4

[41] Leontiou AA, Ladavos AK, Pomonis PJ. Catalytic NO reduction with CO on $La_{1-x}Sr_x(Fe^{3+}/Fe^{4+})O_{3\pm\delta}$ perovskite-type mixed oxides (x = 0.00, 0.15, 0.30, 0.40, 0.60, 0.70, 0.80, and 0.90). Applied Catalysis A: General. 2003;**241**(1):133-141. DOI: 10.1016/S0926-860X(02)00457-X

[42] Wang Y, Cui X, Li Y, Shu Z, Chen H, Shi J. A simple co-nanocasting method to synthesize high surface area

mesoporous $LaCoO_3$ oxides for CO and NO oxidations. Microporous and Mesoporous Materials. 2013;**176**:8-15. DOI: 10.1016/j.micromeso.2013.03.033

[43] Peter SD, Garbowski E, Perrichon V, Primet M. NO reduction by CO over aluminate-supported perovskites. Catalysis Letters. 2000;**70**(1):27-33. DOI: 10.1023/A:1019027619209

[44] Shen S, Weng H. Comparative study of catalytic reduction of nitric oxide with carbon monoxide over the $La_{1-x}Sr_xBO_3$ (B = Mn, Fe, Co, Ni) catalysts. Industrial and Engineering Chemistry Research. 1998;**37**(7):2654-2661. DOI: 10.1021/ie970691g

[45] Varma S, Wani BN, Gupta NM. Redox behavior and catalytic activity of La–Fe–V–O mixed oxides. Applied Catalysis A: General. 2003;**241**(1):341-348. DOI: 10.1016/S0926-860X(02)00492-1

[46] De-La-Torre U, Pereda-Ayo B, Gutiérrez-OrtizMA, González-MarcosJA, González-Velasco JR. Steady-state NH_3-SCR global model and kinetic parameter estimation for NO_x removal in diesel engine exhaust aftertreatment with Cu/chabazite. Catalysis Today. 2017;**296**:95-104. DOI: 10.1016/j.cattod.2017.04.011

[47] Cavataio G, Girard J, Eli Patterson J, Montreuil C, Cheng Y, Lambert C. Laboratory testing of urea-SCR formulations to meet tier 2 bin 5 emissions, SAE Technical Paper 2007-01-1575. 2007. DOI: 10.4271/2007-01-1575

[48] Koebel M, Elsener M, Marti T. NO_x-reduction in diesel exhaust gas with urea and selective catalytic reduction. Combustion Science and Technology. 1996;**121**(1-6):85-102. DOI: 10.1080/00102209608935588

[49] Nova I, Tronconi E, editors. Urea-SCR Technology for $DeNO_x$ After Treatment of Diesel Exhausts. Berlin: Springer; 2014. 715p. DOI: 10.1080/00102209608935588

[50] Urrutxua M, Pereda-Ayo B, Trandafilovic LV, Olsson L, González-Velasco JR. Influence of H_2, CO, C_3H_6, and C_7H_8 as reductants on $DeNO_x$ behavior of dual monoliths for NO_x storage/reduction coupled with selective catalytic reduction. Industrial and Engineering Chemistry Research. 2019;**58**(17):7001-7013. DOI: 10.1021/acs.iecr.9b00149

[51] De La Torre U, Pereda-Ayo B, González-Velasco JR. Cu-zeolite NH_3-SCR catalysts for NO_x removal in the combined NSR–SCR technology. Chemical Engineering Journal. 2012;**207-208**:10-17. DOI: 10.1016/j.cej.2012.06.092

[52] Zhu Z, Liu Z, Niu H, Liu S. Promoting effect of SO_2 on activated carbon-supported vanadia catalyst for NO reduction by NH_3 at low temperatures. Journal of Catalysis. 1999;**187**(1):245-248. DOI: 10.1006/jcat.1999.2605

[53] Tang X, Hao J, Xu W, Li J. Low temperature selective catalytic reduction of NO_x with NH_3 over amorphous MnO_x catalysts prepared by three methods. Catalysis Communications. 2007;**8**(3):329-334. DOI: 10.1016/j.catcom.2006.06.025

[54] Singoredjo L, Korver R, Kapteijn F, Moulijn J. Alumina supported manganese oxides for the low-temperature selective catalytic reduction of nitric oxide with ammonia. Applied Catalysis B: Environmental. 1992;**1**(4):297-316. DOI: 10.1016/0926-3373(92)80055-5

[55] Chmielarz L, Jabłońska M, Strumiński A, Piwowarska Z, Węgrzyn A, Witkowski S, et al. Selective catalytic oxidation of ammonia to nitrogen over Mg-Al, Cu-Mg-Al and Fe-Mg-Al mixed metal oxides doped with noble metals.

Applied Catalysis B: Environmental. 2013;**130-131**:152-162. DOI: 10.1016/j.apcatb.2012.11.004

[56] Zhang R, Luo N, Yang W, Liu N, Chen B. Low-temperature selective catalytic reduction of NO with NH_3 using perovskite-type oxides as the novel catalysts. Journal of Molecular Catalysis A: Chemical. 2013;**371**:86-93. DOI: 10.1016/j.molcata.2013.01.018

[57] Zhang R, Yang W, Luo N, Li P, Lei Z, Chen B. Low-temperature NH_3-SCR of NO by lanthanum manganite perovskites: Effect of A-/B-site substitution and TiO_2/CeO_2 support. Applied Catalysis B: Environmental. 2014;**146**:94-104. DOI: 10.1016/j.apcatb.2013.04.047

[58] Zhang Y, Wang D, Wang J, Chen Q, Zhang Z, Pan X, et al. $BiMnO_3$ perovskite catalyst for selective catalytic reduction of NO with NH_3 at low temperature. Chinese Journal of Catalysis. 2012;**33**(9):1448-1454. DOI: 10.1016/S1872-2067(11)60439-7

[59] Shi H, Li X, Xia J, Lu X, Zuo S, Luo S, et al. Sol-gel synthesis of $LaBO_3$/attapulgite (B=Mn, Fe, Co, Ni) nanocomposite for NH_3-SCR of NO at low temperature. Journal of Inorganic and Organometallic Polymers. 2017;**27**(1):166-172. DOI: 10.1007/s10904-017-0683-9

[60] Wallin M, Cruise N, Klement U, Palmqvist A, Skoglundh M. Preparation of Mn, Fe and Co based perovskite catalysts using microemulsions. Colloids and Surfaces A: Physicochemical and Engineering Aspects. 2004;**238**(1):27-35. DOI: 10.1016/j.colsurfa.2004.02.019

[61] Takahashi N, Shinjoh H, Iijima T, Suzuki T, Yamazaki K, Yokota K, et al. The new concept 3-way catalyst for automotive lean-burn engine: NO_x storage and reduction catalyst. Catalysis Today. 1996;**27**(1):63-69. DOI: 10.1016/0920-5861(95)00173-5

[62] Nova I, Castoldi L, Lietti L, Tronconi E, Forzatti P. On the dynamic behavior of NO_x storage/reduction Pt–Ba/Al_2O_3 catalyst. Catalysis Today. 2002;**75**(1):431-437. DOI: 10.1016/S0920-5861(02)00093-7

[63] Nova I, Castoldi L, Lietti L, Tronconi E, Forzatti P, Prinetto F, Ghiotti G. NO_x adsorption study over Pt–Ba/alumina catalysts: FT-IR and pulse experiments. Journal of Catalysis. 2004;**222**(2):377-388. DOI: 10.1016/j.jcat.2003.11.013

[64] Lietti L, Forzatti P, Nova I, Tronconi E. NO_x storage reduction over Pt-Ba/γ-Al_2O_3 catalyst. Journal of Catalysis. 2001;**204**(1):175-191. DOI: 10.1006/jcat.2001.3370

[65] Lietti L, Nova I, Forzatti P. Role of ammonia in the reduction by hydrogen of NO_x stored over Pt–Ba/Al_2O_3 lean NO_x trap catalysts. Journal of Catalysis. 2008;**257**(2):270-282. DOI: 10.1016/j.jcat.2008.05.005

[66] Onrubia-Calvo JA. Perovskites as alternative for NO_x storage and reduction systems: formulations, mechanism and optimal control [thesis]. Leioa: Universidad del País Vasco; 2019

[67] You R, Meng M, Zhang J, Zheng L, Hu T, Li X. A noble-metal-free SCR-LNT coupled catalytic system used for high-concentration NO_x reduction under lean-burn condition. Catalysis Today. 2019;**327**:347-356. DOI: 10.1016/j.cattod.2018.03.022

[68] Epling WS, Campbell LE, Yezerets A, Currier NW, Parks JE. Overview of the fundamental reactions and degradation mechanisms of NO_x storage/reduction catalysts. Catalysis Reviews. 2004;**46**(2):163-245. DOI: 10.1081/CR-200031932

[69] Pereda-Ayo B, De La Torre U, González-Marcos MP,

González-Velasco JR. Influence of ceria loading on the NO_x storage and reduction performance of model Pt–Ba/Al_2O_3 NSR catalyst. Catalysis Today. 2015;**241**:133-142. DOI: 10.1016/j.cattod.2014.03.044

[70] Chen J, Shen M, Wang X, Qi G, Wang J, Li W. The influence of nonstoichiometry on $LaMnO_3$ perovskite for catalytic NO oxidation. Applied Catalysis B: Environmental. 2013;**134-135**:251-257. DOI: 10.1016/j.apcatb.2013.01.027

[71] Chen J, Shen M, Wang X, Wang J, Su Y, Zhao Z. Catalytic performance of NO oxidation over $LaMeO_3$ (Me=Mn, Fe, Co) perovskite prepared by the sol–gel method. Catalysis Communications. 2013;**37**:105-108. DOI: 10.1016/j.catcom.2013.03.039

[72] Choi SO, Penninger M, Kim CH, Schneider WF, Thompson LT. Experimental and computational investigation of effect of Sr on NO oxidation and oxygen exchange for $La_{1-x}Sr_xCoO_3$ perovskite catalysts. ACS Catalysis. 2013;**3**(12):2719-2728. DOI: 10.1021/cs400522r

[73] Mars P, van Krevelen DW. Oxidations carried out by means of vanadium oxide catalysts. Chemical Engineering Science. 1954;**3**:41-59. DOI: 10.1016/S0009-2509(54)80005-4

[74] Idriss H, Barteau MA. Active sites on oxides: From single crystals to catalysts. Advances in Catalysis. 2000;**45**:261-331. DOI: 10.1016/S0360-0564(02)45016-X

[75] Białobok B, Trawczyński J, Miśta W, Zawadzki M. Ethanol combustion over strontium- and cerium-doped $LaCoO_3$ catalysts. Applied Catalysis B: Environmental. 2007;**72**(3-4):395-403. DOI: 10.1016/j.apcatb.2006.12.006

[76] Dow W, Huang T. Yttria-stabilized zirconia supported copper oxide catalyst: II. Effect of oxygen vacancy of support on catalytic activity for CO oxidation. Journal of Catalysis. 1996;**160**(2):171-182. DOI: 10.1006/jcat.1996.0136

[77] Islam MS, Cherry M, Winch LJ. Defect chemistry of $LaBO_3$ (B = Al, Mn or Co) perovskite-type oxides. Relevance to catalytic and transport behaviour. Journal of the Chemical Society, Faraday Transactions. 1996;**92**(3):479-482. DOI: 10.1039/FT9969200479

[78] Chien C, Shi J, Huang T. Effect of oxygen vacancy on CO-NO-O_2 reaction over yttria-stabilized zirconia-supported copper oxide catalyst. Industrial and Engineering Chemistry Research. 1997;**36**(5):1544-1551. DOI: 10.1006/jcat.1996.0136

[79] Over H. Surface chemistry of ruthenium dioxide in heterogeneous catalysis and electrocatalysis: From fundamental to applied research. Chemical Reviews. 2012;**112**(6):3356-3426. DOI: 10.1021/cr200247n

[80] Constantinou C, Li W, Qi G, Epling WS. NO_X storage and reduction over a perovskite-based lean NO_X trap catalyst. Applied Catalysis B: Environmental. 2013;**134-135**:66-74. DOI: 10.1016/j.apcatb.2012.12.034

[81] Qi G, Li W. Pt-free, $LaMnO_3$ based lean NO_x trap catalysts. Catalysis Today. 2012;**184**(1):72-77. DOI: 10.1016/j.cattod.2011.11.012

[82] He X, Meng M, He J, Zou Z, Li X, Li Z, et al. A potential substitution of noble metal Pt by perovskite $LaCoO_3$ in $ZrTiO_4$ supported lean-burn NO_x trap catalysts. Catalysis Communications. 2010;**12**(3):165-168. DOI: 10.1016/j.catcom.2010.09.016

[83] You R, Zhang Y, Liu D, Meng M, Jiang Z, Zhang S, et al. A series of ceria supported lean-burn NO_x trap catalysts $LaCoO_3/K_2CO_3/CeO_2$ using perovskite

as active component. Chemical Engineering Journal. 2015;**260**:357-367. DOI: 10.1016/j.cej.2014.09.016

[84] You R, Zhang Y, Liu D, Meng M, Zheng L, Zhang J, et al. YCeZrO ternary oxide solid solution supported nonplatinic lean-burn NO_x trap catalysts using $LaCoO_3$ perovskite as active phase. Journal of Physical Chemistry C. 2014;**118**(44):25403-25420. DOI: 10.1021/jp505601x

[85] Onrubia-Calvo JA, Pereda-Ayo B, De-La-Torre U, González-Velasco JR. Strontium doping and impregnation onto alumina improve the NO_x storage and reduction capacity of $LaCoO_3$ perovskites. Catalysis Today. 2019;**333**:208-218. DOI: 10.1016/j.cattod.2018.12.031

[86] Li XG, Dong YH, Xian H, Hernández WY, Meng M, Zou HH, et al. De-NO_x in alternative lean/rich atmospheres on $La_{1-x}Sr_xCoO_3$ perovskites. Energy & Environmental Science. 2011;**4**(9):3351-3354. DOI: 10.1039/C1EE01726H

[87] Xian H, Zhang X, Li X, Li L, Zou H, Meng M, et al. $BaFeO_{3-x}$ perovskite: An efficient NO_x absorber with a high sulfur tolerance. Journal of Physical Chemistry C. 2010;**114**(27):11844-11852. DOI: 10.1021/jp100197c

[88] Dacquin JP, Lancelot C, Dujardin C, Da Costa P, Djega-Mariadassou G, Beaunier P, et al. Influence of preparation methods of $LaCoO_3$ on the catalytic performances in the decomposition of N_2O. Applied Catalysis B: Environmental. 2009;**91**(3-4):596-604. DOI: 10.1016/j.apcatb.2009.06.032

[89] Giraudon JM, Elhachimi A, Wyrwalski F, Siffert S, Aboukaïs A, Lamonier JF, et al. Studies of the activation process over Pd perovskite-type oxides used for catalytic oxidation of toluene. Applied Catalysis B: Environmental. 2007;**75**(3):157-166. DOI: 10.1016/j.apcatb.2007.04.005

[90] Say Z, Dogac M, Vovk EI, Kalay YE, Kim CH, Li W, et al. Palladium doped perovskite-based NO oxidation catalysts: The role of Pd and B-sites for NO_x adsorption behavior via in-situ spectroscopy. Applied Catalysis B: Environmental. 2014;**154-155**:51-61. DOI: 10.1016/j.apcatb.2014.01.038

[91] Zhang R, Alamdari H, Kaliaguine S. Fe-based perovskites substituted by copper and palladium for NO+CO reaction. Journal of Catalysis. 2006;**242**(2):241-253. DOI: 10.1016/j.jcat.2006.05.033

[92] Yoon DY, Kim YJ, Lim JH, Cho BK, Hong SB, Nam I, et al. Thermal stability of Pd-containing $LaAlO_3$ perovskite as a modern TWC. Journal of Catalysis. 2015;**330**:71-83. DOI: 10.1016/j.jcat.2015.07.013

[93] Li C, Wang C, Lin Y. Pd-integrated lanthanum-transition metal perovskites for methanol partial oxidation. Catalysis Today. 2011;**174**(1):135-140. DOI: 10.1016/j.cattod.2011.01.038

[94] Nishihata Y, Mizuki J, Akao T, Tanaka H, Uenishi M, Kimura M, et al. Self-regeneration of a Pd-perovskite catalyst for automotive emissions control. Nature. 2002;**418**:164. DOI: 10.1038/nature00893

[95] Zhou K, Chen H, Tian Q, Hao Z, Shen D. Xu X. Pd-containing perovskite-type oxides used for three-way catalysts. Journal of Molecular Catalysis A: Chemical. 2002;**189**(2):225-232. DOI: 10.1016/S1381-1169(02)00177-2

[96] Rodríguez GCM, Kelm K, Heikens S, Grünert W, Saruhan B. Pd-integrated perovskites for TWC applications: Synthesis, microstructure and N_2O-sele. Catalysis Today. 2012;**184**(1):184-191. DOI: 10.1016/j.cattod.2011.12.026

[97] Zhao D, Gao Z, Xian H, Xing L, Yang Y, Tian Y, et al. Addition of Pd on $La_{0.7}Sr_{0.3}CoO_3$ perovskite to enhance catalytic removal of NO_x. Industrial and Engineering Chemistry Research. 2018;**57**(2):521-531. DOI: 10.1021/acs.iecr.7b04399

[98] Ye J, Yu Y, Meng M, Jiang Z, Ding T, Zhang S, et al. Highly efficient NO_x purification in alternating lean/rich atmospheres over non-platinic mesoporous perovskite-based catalyst $K/LaCoO_3$. Catalysis Science & Technology. 2013;**3**(8):1915-1918. DOI: 10.1039/C3CY00155E

[99] Wen W, Wang X, Jin S, Wang R. $LaCoO_3$ perovskite in $Pt/LaCoO_3/K/Al_2O_3$ for the improvement of NO_x storage and reduction performances. RSC Advances. 2016;**6**(78):74046-74052. DOI: 10.1039/C6RA18273A

[100] Ueda A, Yamada Y, Katsuki M, Kiyobayashi T, Xu Q, Kuriyama N. Perovskite catalyst (La, Ba)(Fe, Nb, Pd)O_3 applicable to NO_x storage and reduction system. Catalysis Communications. 2009;**11**(1):34-37. DOI: 10.1016/j.catcom.2009.08.008

[101] Hodjati S, Petit C, Pitchon V, Kiennemann A. Absorption/desorption of NO_x process on perovskites: Impact of SO_2 on the storage capacity of $BaSnO_3$ and strategy to develop thioresistance. Applied Catalysis B: Environmental. 2001;**30**(3):247-257. DOI: 10.1016/S0926-3373(00)00249-6

[102] Hodjati S, Vaezzadeh K, Petit C, Pitchon V, Kiennemann A. Absorption/desorption of NO_x process on perovskites: Performances to remove NO_x from a lean exhaust gas. Applied Catalysis B: Environmental. 2000;**26**(1):5-16. DOI: 10.1016/S0926-3373(99)00143-5

[103] Xian H, Zhang X, Li X, Zou H, Meng M, Zou Z, et al. Effect of the calcination conditions on the NO_x storage behavior of the perovskite $BaFeO_{3-x}$ catalysts. Catalysis Today. 2010;**158**(3):215-219. DOI: 10.1016/j.cattod.2010.03.026

[104] Xian H, Li F, Li X, Zhang X, Meng M, Zhang T, et al. Influence of preparation conditions to structure property, NO_x and SO_2 sorption behavior of the $BaFeO_{3-x}$ perovskite catalyst. Fuel Processing Technology. 2011;**92**(9):1718-1724. DOI: 10.1016/j.fuproc.2011.04.021

[105] Li X, Chen C, Liu C, Xian H, Guo L, Lv J, et al. Pd-doped perovskite: An effective catalyst for removal of NO_x from lean-burn exhausts with high sulfur resistance. ACS Catalysis. 2013;**3**(6):1071-1075. DOI: 10.1021/cs400136t

[106] Ma A, Wang S, Liu C, Xian H, Ding Q, Guo L, et al. Effects of Fe dopants and residual carbonates on the catalytic activities of the perovskite-type $La_{0.7}Sr_{0.3}Co_{1-x}Fe_xO_3$ NO_x storage catalyst. Applied Catalysis B: Environmental. 2014;**146**:24-34. DOI: 10.1016/j.apcatb.2013.06.005

[107] Wang X, Qi X, Chen Z, Jiang L, Wang R, Wei K. Studies on SO_2 tolerance and regeneration over perovskite-type $LaCo_{1-x}Pt_xO_3$ in NO_x storage and reduction. Journal of Physical Chemistry C. 2014;**118**(25):13743-13751. DOI: 10.1021/jp5044255

[108] Minami, Toshitake (Kanagawa, JP) 2009. Device for purifying exhaust gas of a diesel engine United States 20090250041 http://www.freepatentsonline.com/y2009/0250041.html

[109] Sakurai K. Exhaust purifying system for internal combustion engine. US Patent Appl. 2011/0138783 A1 (16-06-2011)

[110] De La Torre U, Urrutxua M, Pereda-Ayo B, González-Velasco JR.

On the Cu species in Cu/beta catalysts related to $DeNO_x$ performance of coupled NSR-SCR technology using sequential monoliths and dual-layer monolithic catalysts. Catalysis Today. 2016;**273**:72-82. DOI: 10.1016/j.cattod.2016.02.054

[111] De-La-Torre U, Pereda-Ayo B, Moliner M, González-Velasco JR, Corma A. Cu-zeolite catalysts for NO_x removal by selective catalytic reduction with NH_3 and coupled to NO storage/reduction monolith in diesel engine exhaust after treatment systems. Applied Catalysis B: Environmental. 2016;**187**:419-427

[112] Pereda-Ayo B, Duraiswami D, Delgado JJ, López-Fonseca R, Calvino JJ, Bernal S, et al. Tuning operational conditions for efficient NO_x storage and reduction over a Pt–Ba/Al_2O_3 monolith catalyst. Applied Catalysis B: Environmental. 2010;**96**(3-4):329-337

[113] Pereda-Ayo B, Duraiswami D, González-Marcos JA, González-Velasco JR. Performance of NO_x storage-reduction catalyst in the temperature-reductant concentration domain by response surface methodology. Chemical Engineering Journal. 2011;**169**(1-3):58-67. DOI: 10.1016/j.cej.2011.02.052

[114] Pereda-Ayo B, Duraiswami D, González-Velasco JR. Control of NO_x storage and reduction in NSR bed for designing combined NSR–SCR systems. Catalysis Today. 2011;**172**(1):66-72. DOI: 10.1016/j.cattod.2011.01.043

[115] Cumaranatunge L, Mulla SS, Yezerets A, Currier NW, Delgass WN, Ribeiro FH. Ammonia is a hydrogen carrier in the regeneration of Pt/BaO/Al_2O_3 NO_x traps with H_2. Journal of Catalysis. 2007;**246**(1):29-34. DOI: 10.1016/j.jcat.2006.11.008

[116] Pereda-Ayo B, González-Velasco JR, Burch R, Hardacre C, Chansai S. Regeneration mechanism of a lean NO_x trap (LNT) catalyst in the presence of NO investigated using isotope labelling techniques. Journal of Catalysis. 2012;**285**(1):177-186. DOI: 10.1016/j.jcat.2011.09.028

Chapter 6

Perovskite-Based Materials for Energy Applications

Mirela Dragan, Stanica Enache, Mihai Varlam and Konstantin Petrov

Abstract

The role of energy in modern society is fundamental. Constraints due to the emissions of air pollutants from the excessive use of fossil fuels have increased dramatically in the last years. Over the years various devices and systems have been developed to transform energy from forms supplied by nature to forms that can be used by people. Another issue is to absorb energy generated at one time and to discharge it to supply power at a later time, what is called energy storage. This is also a matter to focus when it comes to searching for solutions of energy problems. Perovskites are promising key materials for energy applications, and in this chapter is a literature review summarizing the reported progress in energy applications of perovskite-type ceramic materials. To understand the fundamental nature of structure–property relationships, defect chemistry plays an important role. This paper, a mini-review, briefly describes from available literature and summarizes accordingly. It is focused on perovskite crystal structures, perovskite materials for solid oxide fuel cells, perovskite electrocatalyst and photocatalysts, and perovskite transport features.

Keywords: perovskite, perovskite crystal structure, defect chemistry, perovskite synthesis, transport properties

1. Introduction

There is going to be huge demand for energy. The growing population and the growing of industrialization will increase the demand for energy. The world needs *more energy* to allow global living standards to continue to improve.

Currently most of our energy comes from fossil fuels, which originated from deep within the Earth's crust. This had disastrous effects on the planet because the burning of coal, oil, and gas has been linked to the rising levels of greenhouse gases on the Earth's atmosphere, generating climate change. The global energy landscape requires improvements. There is a slow going on transition to a more sustainable energy system. Not only to meet the need of rising energy demand but also in terms of policy, reducing carbon emission energy systems is the biggest challenges of our time.

Along with these challenges come opportunities—and that is what makes this a really exciting time for material science with respect to efficient and cost-effective energy conversion and energy storage. The process of changing energy from one

IntechOpen

form to another is energy conversion, and energy storage is the capture of energy produced at one time for use at a later time.

This work aims to provide an overview, based on available literature, of perovskite materials for energy applications and will focus especially on solid oxide fuel cells for efficient power generation from fuels and electrocatalysts for oxygen reduction reaction, oxygen evolution reaction, as well as on the defect chemistry in general of such materials. A large number of articles related on this topic have been published in the past decades. Unfortunately, it is not possible to cover all the aspects and references in the literature.

Due to their properties, the perovskite materials are of considerable technological importance covering a very broad range of practical applications. Notable is the discovery in the 1940s of the ferroelectric properties of barium titanate ($BaTiO_3$) used in electronics for capacitors and transducers [1]. In the mid-1980s, the first high-temperature superconductor was discovered, lanthanum barium copper oxide and, in 1987 Nobel Prize in physics, was awarded for this discovery [2]. As of 2012, perovskites have been identified as possible inexpensive base materials for high-efficiency commercial photovoltaics, and perovskites also have optoelectronic properties such as strong light absorption and facilitated charge transport [3]. Some of perovskites' typical properties are ferromagnetism [4], piezoelectricity [5, 6], electrical conductivity [7, 8], superconductivity [9, 10], ion conductivity [11, 12], magnetism [13, 14], catalytic properties [15, 16], electrode materials [17, 18], and optical [19, 20].

This work will describe their defect chemistry which plays an important role in energy applications where the transport properties are the main players. Their defect chemistry is responsible for properties like ionic conductivity [21], mixed conductivity [22], proton conductivity [23], and catalytic conductivity [24] which make perovskite being used for solid oxide fuel cells (SOFC), electrolyte, SOFC electrode, and catalyst.

2. Perovskites and related structures

The perovskite structure is adopted by many compounds residing on the generic formula ABX_3, the same type of crystal structure as calcium titanium oxide ($CaTiO_3$), constituting the family of perovskite compounds. We owe the discovery of perovskite to Gustav Rose, a German mineralogist who performed the studies in

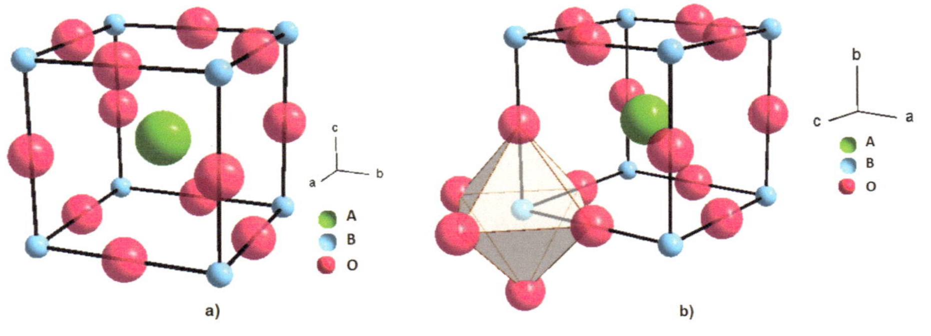

Figure 1.
Cubic perovskite unit cell. Green spheres represent A cations, blue spheres represent B cations, and red spheres represent oxygen anions forming an octahedron. (a). In the perovskite structure with the large cation at the cube center, the small cations are on the corners, and the anions are located at the midpoint of each edge. (b). It emphasizes the octahedral coordination of the small cation.

the Ural Mountains in Russia back in 1893. During that time, he identified perovskite, a naturally occurring oxide species, with the chemical formula $CaTiO_3$ and named it after Russian mineralogist Count Lev Alekseyevich von Perovski [25].

The ideal perovskite-type structure is cubic with space group Pm3m [26]. The positive charge B-type cations are occupying the centers of corner-shared octahedra of negative charge X-type anions such as oxygen halides, sulfides, or nitrides, and positive charge A-type cations are filling the resulting interstices. We restrict this

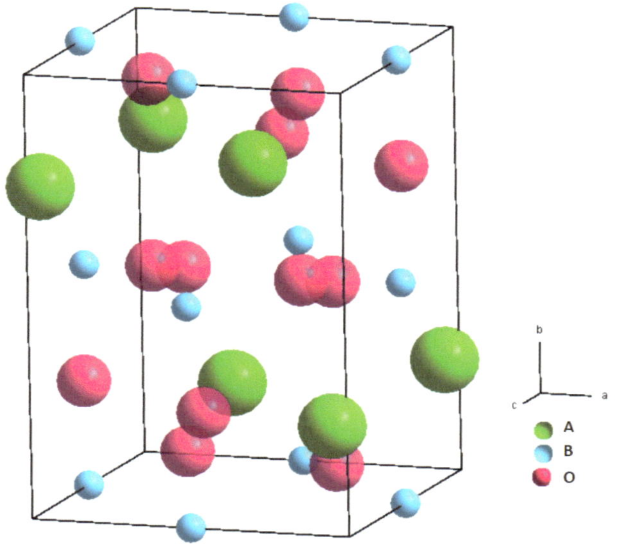

Figure 2.
The orthorhombic phase of perovskite. Green spheres represent A cations, blue spheres represent B cations, and red spheres represent oxygen anions.

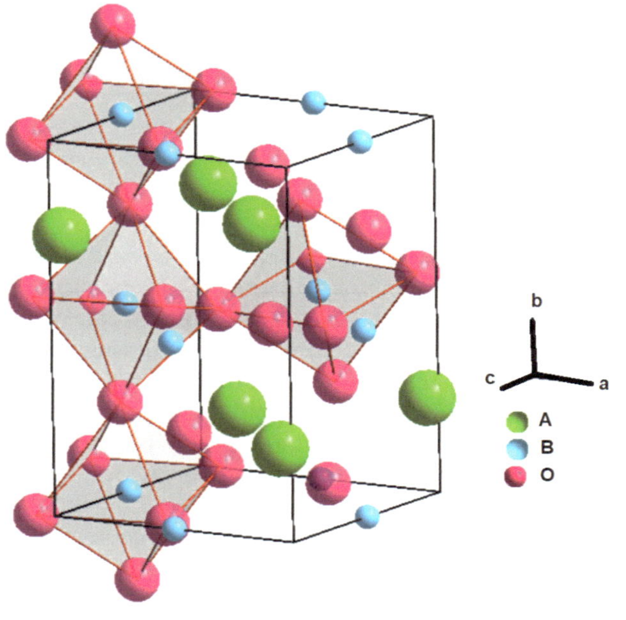

Figure 3.
The tetragonal phase of perovskite. Green spheres represent A cations, blue spheres represent B cations, and red spheres represent oxygen anions.

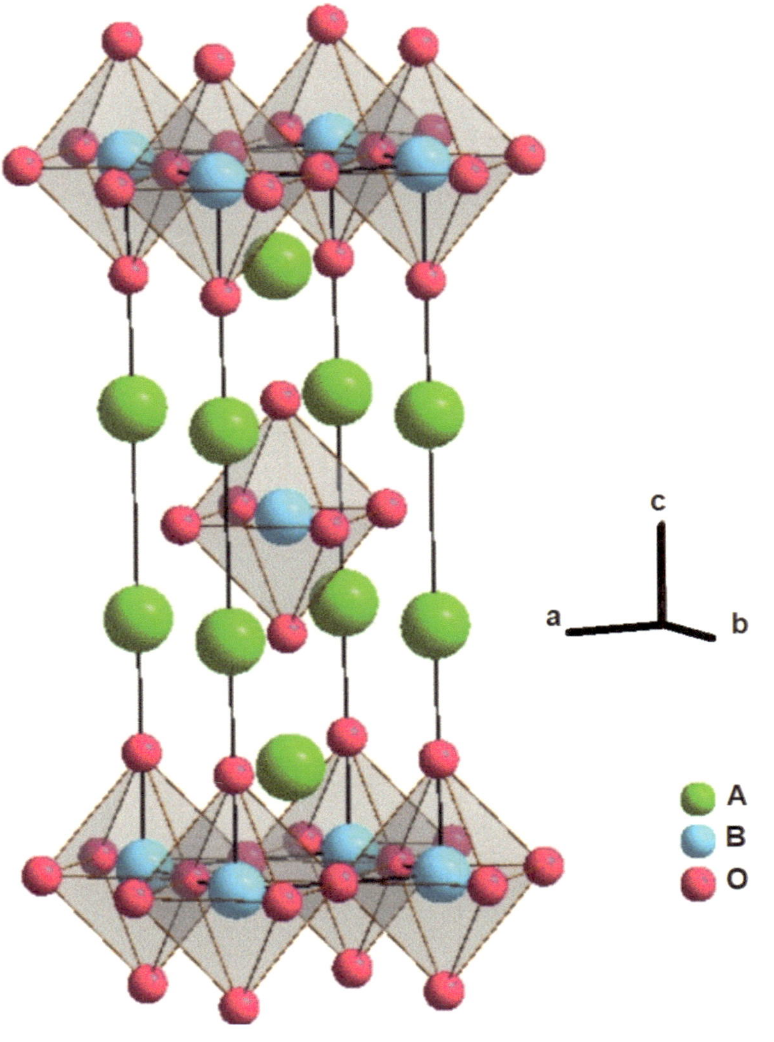

Figure 4.
The Ruddlesden-Popper phases. Green spheres represent A cations, blue spheres represent B cations, and red spheres represent oxygen anions.

study to the oxide perovskites. In **Figure 1a**, alternative ways to view the perovskite structure are displayed: the cubic perovskite unit cell. **Figure 1b** emphasizes the octahedral coordination of the small cation. Green spheres represent A cations, blue spheres represent B cations, and red spheres represent oxygen anions forming an octahedron. Most perovskites are distorted and do not have the ideal undistorted cubic structure. Few perovskite compounds actually form the ideal cubic structure. The mineral perovskite itself, calcium titanate $CaTiO_3$, is an orthorhombic distortion of the basic structure; strontium titanate $SrTiO_3$ is often used as the prototype.

The equation determined by Goldschmidt correlates geometrically crystal structures in terms of the ionic packing using the Goldschmidt's tolerance factor t [27]. Favorable for cubic perovskite structure is the tolerance factor having values between 0.8 and 1; otherwise the ideal cubic structure is a distorted structure. Empirical mathematical expression involving the unit cell length ratio, here r_A, r_B, and r_O, is the ionic radii for A, B, and O, respectively; t is given as

$$t = \frac{(r_A + r_O)}{\sqrt{2}(r_B + r_O)} \tag{1}$$

Cations with large ionic radius are occupying A sites, and cations with smaller ionic radius are occupying B sites. A and O form cubic closest packing, and B is in the octahedral voids in the packing. In an ideal structure, where the atoms are

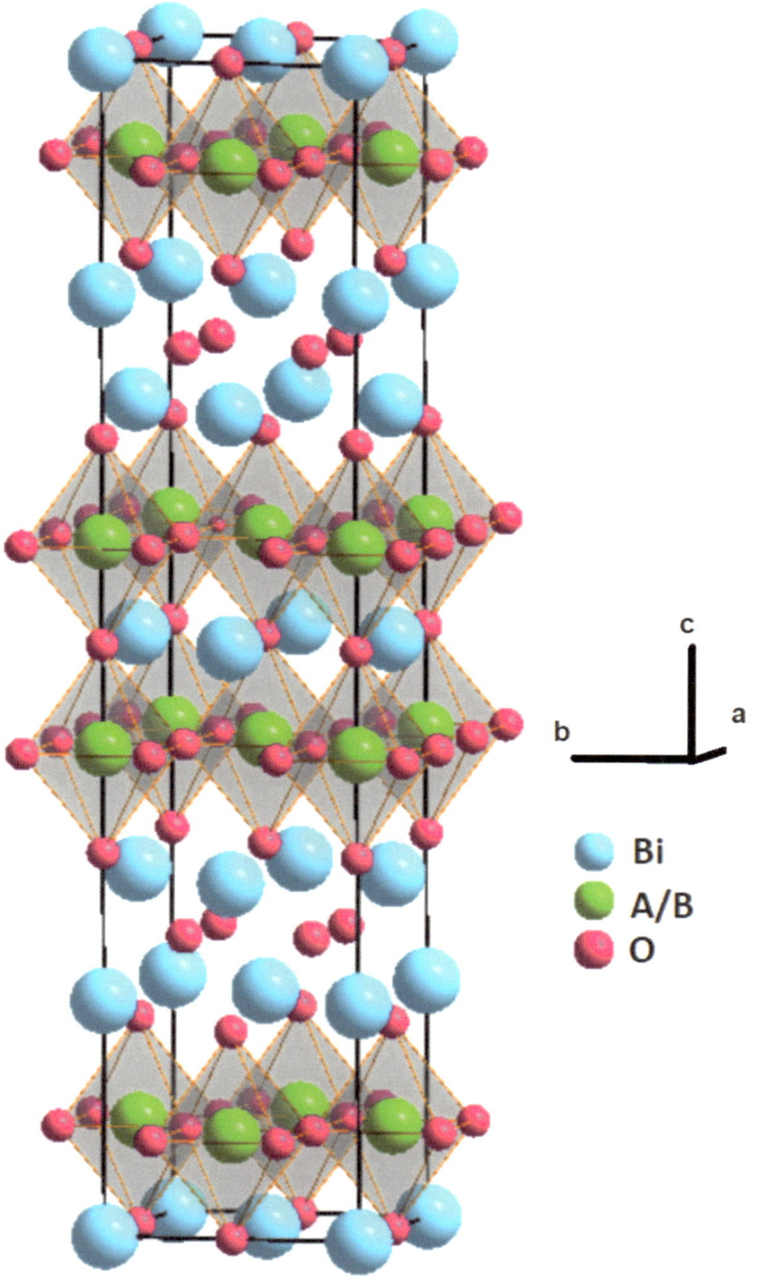

Figure 5.
The Aurivillius n = 1 structure. Green spheres represent A cations, blue spheres represent B cations, and red spheres represent oxygen anions.

simply bonding to one another, the B-O distance is equal to a/2, whereas the A-O distance is equal to $(a/\sqrt{2})$ where a = length of unit cell.

The orthorhombic (**Figure 2**) and tetragonal (**Figure 3**) phases are the most common non-cubic variants [26] with octahedral distortions and translational A site offsets. As previously mentioned, green spheres represent A cations, blue spheres represent B cations, and red spheres represent oxygen anions.

Perovskites allow chemical tailoring because of the wide range of ions and valences which they can accommodate. By suitable formulation many valuable properties can be tailored. One interesting quality of perovskites is that they can accommodate more than one element for both A- and B sites constituting complex perovskites. These can be represented as $(A^1{}_{1-x}A^2{}_x)BO_3$, $A(B^1{}_{1-x}B^2{}_x)O_3$. Substitutional aliovalent cations residing on the host cation sites generate oxygen vacancies. Furthermore for the B site, the element of transition metal can be in different oxidation states $A(B'_xB''_y)O_3$, where x + y = 1. Aliovalent substitutions require a charge compensation mechanism because the ionic compound must be neutral. Therefore, either ion vacancies are created or one of the metals is partially or fully reduced or oxidized.

Another interesting and useful to deal with is the complex set of phase relationships. These are useful in such materials because they are sensitive to external conditions, including temperature, pressure, strain, and composition [28].

The perovskite structure also lends itself to the building of superstructures as seen in **Figure 4**. The Ruddlesden-Popper phases are a series of structures consisting of sequences of max 3 n; perovskite blocks are separated by a rock-salt structure block. The first member of the Ruddlesden-Popper series is A_2BO_4 or ABO_3-AO which is isostructural with potassium tetrafluoridenickelate (II) K_2NiF_4, the prototype structure.

A second family of superstructures are the Aurivillius phases (**Figure 5**) [29]. This is represented by the general formula $(Bi_2O_2)(A_{n-1}B_nO_{3n+1})$, where A is a large monovalent, divalent, or trivalent 12-coordinate cation and B is a small trivalent, pentavalent, or hexavalent metallic 6-coordinate cation. Basically, their structure is built by alternating layers of $[Bi_2O_2]_2{}^+$ and perovskite blocks $(A_{n-1}B_nO3_{n+1})_2$ that contain a layer of octahedral B sites.

Various synthesis methods have been developed for the preparation of perovskite materials while managing efficiently the environmentally friendly processing in conjunction with phase structure tailored to the needed properties. Small changes in the crystal will have a large impact on functional properties, crystallinity, and stability. Among these techniques, dry ways like solid-state reaction mechanochemical processing or wet routes like sol–gel and microwave methods are widely used.

3. Perovskite defect chemistry

In crystalline state an array of atoms is regularly repeated in three dimensions. This construction leads to a perfect crystalline solid with all atoms present and located in their ideal positions. However, a zero entropy is needed for the perfect crystalline state to exist. Above absolute zero temperature is a probability that defects from ideality will exist. All real materials contain defects. These defects, in terms of dimensionality, can be vacancies and interstitials which are point defects, dislocations which are line defects, and grain boundaries which are plane defects.

Point defects cause an increase of the configurational entropy contribution to the free energy at nonzero temperatures and will therefore always be present.

The formation of a point defect can be considered as a chemical reaction. At constant temperature, T, and pressure, p, the reaction is carried on in the direction that lowers the Gibbs free energy as follows:

$$G = U + pV - TS = H - TS \quad (2)$$

where H is the enthalpy, U is the internal energy, S is the entropy, and V is the volume. The enthalpy H is defined as

$$H = U + pV \quad (3)$$

The change in free energy for the formation of n independent defects can be written as follows:

$$\Delta G = n\Delta_f G - T\Delta_f S_c \quad (4)$$

In the above equation, the configurational entropy $\Delta_f S_c$ is the part of the entropy change associated with randomly distributing n defect in the material. For a single defect, the formation free energy $\Delta_f G$ is independent of the number of defects and can be written as follows:

$$\Delta_f G = \Delta_f U + p\Delta_f V - T\Delta_f S \quad (5)$$

The point defects are a nonstoichiometric perturbation of the ideal lattice having or not electrically charge. Materials can also be prepared in such a way as to increase the number of defects.

There are two main categories for point defects: intrinsic defects and extrinsic defects. The intrinsic defects are internal to the crystal. The extrinsic defects are created when an impurity atom or ion is inserted into the lattice. The main categories of intrinsic defects are Schottky disorder and Frenkel disorder.

The Schottky disordering mechanism presumes that atoms leave their sites in the crystal bulk and rebuild the crystal lattice on the surface. As a result, the vacancies are formed in both cation and anion sublattices. According to the Frenkel mechanism, an atom moves from its regular site to the nearest interstitial position; hence, two types of defect are formed in the crystal, namely, the vacancy and the interstitial atom.

A vacancy is formed if an atom is not present on the site that it should occupy in a perfect crystal. By removing one atom from the chemical formula of perovskite, a set of vacancies called Schottky defect is created, which is the removal of an oxygen atom. Oxygen ion has a charged of −2. The defect left behind after removal, a charge of +2, for an oxygen atom will be $V_{\ddot{O}}$ in Kröger-Vink notation. The following equation is written for the creation of oxygen vacancy, Schottky disorder:

$$O_O^X = \frac{1}{2}O_2 + V_{\ddot{O}} + 2e' \quad (6)$$

V indicates a vacancy, the subscript O indicates the oxygen host site, and superscript denoted by two dots is the defect charge of +2.

An interstitial is an ion situated on any site that would be unoccupied in a perfect crystal. The interstitial is denoted by a subscript *i*, in Kröger-Vink notation. Frenkel defects deal with the displacement of the ions from its sites to interstitial sites.

When an ion is substituted with an ion of different valence, another type of point defect is created, called substitutional defect.

Defect chemistry complies with the conservation rules which indicate the charge conservation by maintaining crystal electrical neutrality overall, and the mass balance and host structure should be preserved.

The possible point defects in ABO_3 perovskite are the oxygen vacancies, A site vacancies, B vacancies, A interstitials, and B interstitials. The metal interstitial, B interstitial, defects are not energetically favored and less likely to be present.

Let us have a look at the $LaCoO_3$ simultaneous doping on the A site and B site, with Sr., respectively, Fe and Cr. This is a way to increase oxide ion conductivity and at the same time to retain the required thermodynamic stability of doped lanthanum cobaltite.

Substitution of Cr^{3+} for Co^{3+} in the $LaCoO_3$ results in the formation of Cr^{4+} according to the reaction

$$\mathrm{Co_{Co}^{X}} + \mathrm{Cr_{Co}^{X}} = +\mathrm{Co_{Co}'} + \mathrm{Cr_{Co}^{\bullet}} \tag{7}$$

The charge disproportionation equation for Co site is

$$2\mathrm{Co_{Co}^{X}} = \mathrm{Co_{Co}^{\bullet}} + \mathrm{Co_{Co}'} \tag{8}$$

Using Kröger-Vink notations Co_{Co}^{X}, $Co_{Co}^{\bullet}$, Cr^{4+}, Co_{Co}', Cr_{Co}^{X}, and $Cr_{Co}^{\bullet}$ indicates Cr^{3+}, Co^{4+}, Co^{2+}, and Cr^{3+}, respectively.

For the ions of Fe^{3+} doping on the Co^{3+} site, is a similar behavior. Their interaction is leading to form Fe^{2+}, Fe^{4+} and Co^{2+}, Co^{4+}, respectively.

Sr^{2+} on La^{3+} sites is represented by the negatively effective charged Sr_{La}'.

Within the framework of standard band approach, thermally activated electrons can jump from the valence band via the bandgap toward the conduction band. For conduction band electrons e and valence band holes h, the relevant equation is

$$\mathrm{nil} = \mathrm{e}' + \mathrm{h}^{\bullet} \tag{9}$$

A free electron in the conduction band and an itinerant hole in the valence band appear simultaneously.

Owing to its intrinsic zero-dimensional nature, a point defect, or any single atom, is difficult to observe experimentally, and much of the knowledge about point defects is inferred by implicit methods.

4. Perovskites transport properties

The key important factors for the ionic transport of perovskite materials are their defect chemistry and crystal structure. Point defects, mainly ionic defects, have major impact on transport properties like ionic diffusion and ionic conductivity. These properties are the most important in technical applications mentioned in this chapter, dealing with movement of ions within crystal.

Vacancies, interstitials, and substitutional defects can all be charged. The linear response that relates the current density to the applied field is the conductivity. This proportionality constant can be expressed as follows:

$$\mathrm{I_i} = \sigma_i \mathrm{E} \tag{10}$$

where E is the applied electric field in $V^{*}cm^{-1}$, I_i is the current density for species i in $A^{*}cm^{-2}$, and σ_i is the conductivity σ_i in $1^{*}\Omega^{-1*}cm^{-1}$.

The charge may be transported by electrons or holes, by ions, or by both. There are also solids that exhibit simultaneously significant levels of both ionic and electronic transport and are referred to as mixed conductors, MIECs. While ionic conduction is mainly related to crystal structure, electronic conduction is determined by the electronic bandgap.

The conductivity is proportional to the concentration of charge carriers noted c_i carriers/Volume; the charge they carry is z_ie (C/charge), where e is the unit electronic charge, mobility noted μi, which is their ability to move in an electric field.

Having z_i unit charges on the carrier is described for their mobility to move in an electric field by their mobility μ_i; the conductivity is described by the relations

$$\sigma_i = c_i z_i e \mu_i \tag{11}$$

where c_i is the concentration of charge carriers per volume and z_ie the charge they carry in C per charge.

In response to a concentration gradient, the system attempts to return to a homogenous equilibrium state by eliminating the gradient. The Fick's first law describes the unidirectional diffusion:

$$J_i = D_i \frac{dc_{i-}}{dx} \tag{12}$$

where $dc_i/_{dx}$ is the concentration gradient of species i, J_i is its flux in particles*cm^{-2}*s^{-1}, and D_i is the diffusion constant in particles*cm^{-4}.

5. Clean energy applications

5.1 Perovskites in solid oxide fuel cell

A solid oxide fuel cell, SOFC, essentially consists of two porous electrodes separated by a dense, oxide ion-conducting electrolyte. The operating principle of such a cell is illustrated in **Figure 6**. The oxygen supplied at the cathode or, air electrode, reacts with incoming electrons from the external circuit to form oxide ions, which migrate to the anode or, fuel electrode, through the oxide ion conducting electrolyte. At the anode side, the oxide ions combine with H_2, CO in the fuel to form H_2O, CO_2, with the effect of liberating electrons. Electrons flow from the anode through the external circuit to the cathode. To keep the cell resistance low, the electrolyte is fabricated in the form of a thin film.

Due to its operating conditions there are some limitations to materials used for SOFC.

These materials should be durable without changes of the required properties. Until now, there have been several researches to develop and fabricate materials to meet requirements of SOFC. **Table 1** summarizes such perovskite materials accordingly to the SOFC component.

Electrolyte may carry either oxide ion O^{2-} or proton H^+ and should have high ionic conductivity and uniform features in structure. The important properties of cathodes are high electronic conductivity and thin porous layer where the oxygen reduction reaction takes place. The anode materials must be chemically compatible with electrolyte. Thermal expansion also is a requirement to match with electrolyte thermal expansion characteristics. The thermal expansion of solids depends on their structure symmetry and may be either isotropic or anisotropic. Properties like electrical conductivity, large triple phase boundary, and high electrocatalytic

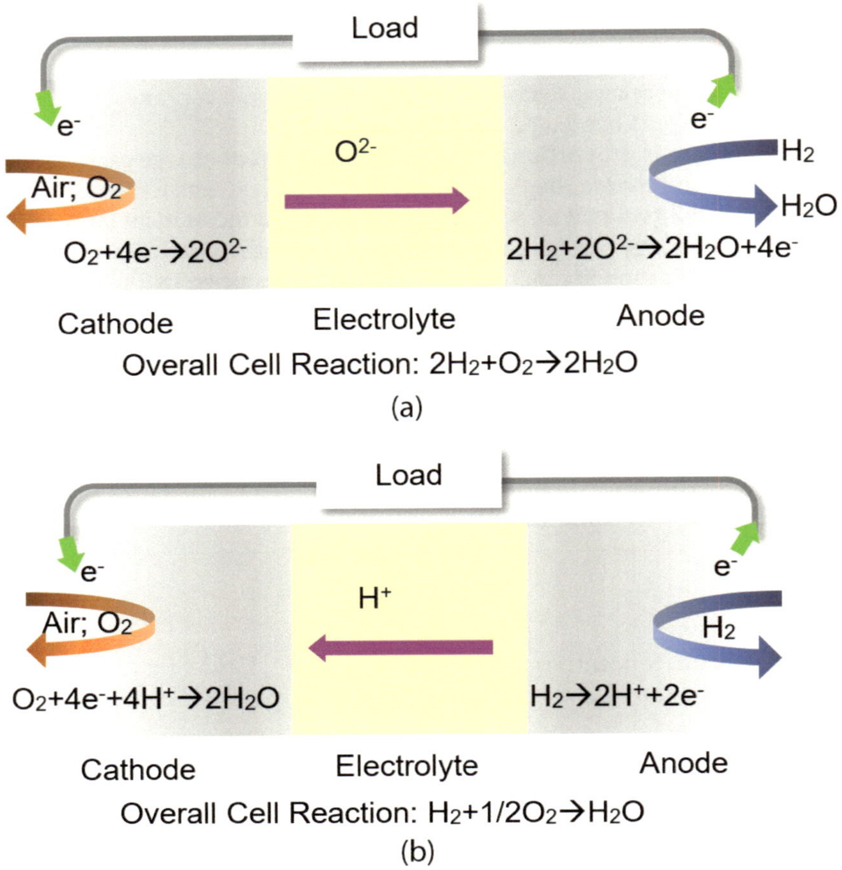

Figure 6.
Operating principle of a solid oxide fuel cell. (a) Operating principle of a solid oxide fuel cell: Oxygen ion-conducting type cell. (b) Operating principle of a solid oxide fuel cell: Hydrogen ion-conducting type cell.

Component part	Material
Electrolyte	$LaGaO_3$-type [30]: (La,Sr)(Ga,Mg)O_3, (La,Sr)(Ga,MgCo)O_3, (La,Sr)(Ga,Mg,Fe) O_3, (La,Sr)(Ga,Mg,Co,Fe)O_3 $LaAlO_3$-type [31]: (La,Ca)AlO_3, (La,Ba)AlO_3 Brownmillerite perovskite: $BaZrO_3$ [32]
Cathode	$LaMnO_3$-type [33]: (La,Sr)MnO_3, (La,Ca)MnO_3 LaCoO3-type [34]: (La,Sr)CoO_3, (La,Ca)CoO_3 (La,Sr)FeO_3 [35] (Sm,Sr)CoO_3, (Sm,Nd)CoO_3, $LaNiO_3$ K_2NiF_4 structure
Anode	$SrNbO_3$ [36]; $SrVO_3$ [37]
Separator	$LaCrO_3$ type [38, 39]: (La,Sr)CrO_3, (La,Ca)CrO_3

Table 1.
Typical materials used in solid oxide fuel cells.

activity provide a mechanism for electronic conductivity, constituting important qualities to take in account from anode materials. Among the materials developed for SOFC components, (La, Sr)MnO_3 in air shows a conductivity of 300 S/cm [40],

(La, Sr)(Co, Fe)O_3 about 330 S/cm [41], and (La, Sr)CoO_3 about 1.22–1.60 S/cm [42, 43]. (Pr, Ba, Sr)(Co, Fe)O_5 is a promising cathode material with power densities about 2.2 W/cm^2 at 600°C and has potential for commercially viable SOFC technologies [44]. The choice of material combinations for the SOFC components is crucial for determining their performances.

5.2 Perovskites in catalysis

Transition metal perovskites are important catalyst materials. The catalytic activity for perovskite materials often resides with metal oxide surface sites, and efficient use of the metals and space available requires small particles, located on a mostly inert support to enhance the thermal stability of the catalyst. Direct electrochemical water splitting is considered a key process in the development of novel energy storage systems, crucial for a sustainable and environmentally friendly energy economy. Water electrolyzers can convert water into hydrogen and oxygen through an electrochemical process, allowing H_2 to be stored as an energy vector as is seen in **Figure 7**.

However, the overpotentials at the anode side where the oxygen evolution reaction, OER, takes place are substantial, even when highly active, precious metal catalysts are used. Therefore, the development of anode materials based on inexpensive and abundant elements, displaying both high OER activity and stability, appears to be a crucial point toward the development of new-generation hydrogen-based storage systems. The values of OER activity about four times higher than that of bulk $LaCoO_3$ compound which was found to be 1.87 A/g were reported for porous and nanosphere $LaCoO_3$. For porous $LaCoO_3$, an OER activity of 7.51 A/g was reported, and for hollow $LaCoO_3$ nanospheres, an OER activity of 12.58 A/g was reported. Both values are at 1.60 V [45].

The oxygen evolution reaction OER in alkaline has the net reaction:

$$4HO^- = 2H_2O + O_2 + 4e^- \tag{13}$$

Typically in electrocatalysis the ion of interest is the B site assumed to be an active site for OER. From a crystal field theory perspective, the octahedrally coordinated B site state d will split into several levels. The states of interest for catalysis will be the antibonding e_g and t_{2g} states since they are typically the occupied states with highest energy and their filling will roughly determine the strength of the B-O bond.

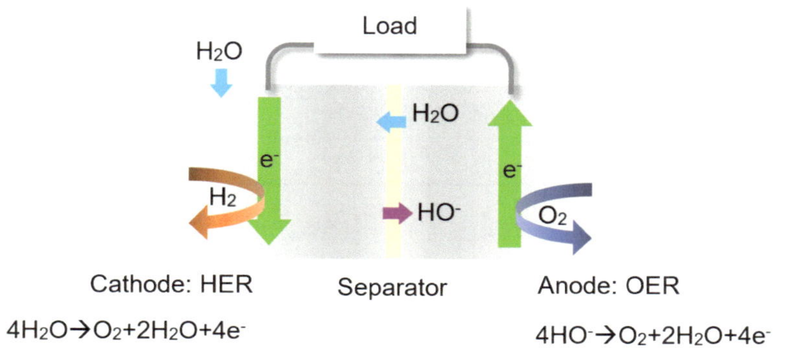

Figure 7.
Working principle of an alkaline electrolysis cell. When the direct current is applied to the water, oxygen and hydrogen are separated from the water. Oxygen arises at the anode while the hydrogen at the cathode.

The OER activities of a series of perovskites were found to form a volcano trend when plotted versus the e_g orbital filling determined using X-ray absorption near-edge structure (XANES) and spin states inferred [46]. Perovskites with an e_g orbital occupancy of approximately 1 were found to be the most active. Based on this principle, highly active perovskite catalysts like $Ba_{0.5}Sr_{0.5}Co_{0.8}Fe_{0.2}O_3$ with a conductivity of 8.58×10^{-5} S/cm, $LaNiO_3$ which has 2.39 S/cm, and $LaCoO_3$ can be used to rationally choose materials as candidates for promising OER catalyst [45, 47–52].

5.3 Perovskites in photocatalysis

In general, an efficient photocatalyst, also called an ideal semiconductor, should include light-harvesting and redox capabilities to facilitate the desired chemical reactions, thus achieving the targeted reaction. Inorganic semiconductor materials should have adequate capability to absorb solar energy across a broad spectrum. Because of their light-harvesting property, potential photocatalysts absorb solar energy, leading to the generation of photoelectrons in the conduction band and holes in the valence band for their possible use in redox reactions.

The redox nature of a photocatalyst as an intrinsic property dictates solar energy conversion efficiency. Most semiconductors include metals/mixed metal oxides used as efficient photocatalysts with the exceptional adaptability of their properties

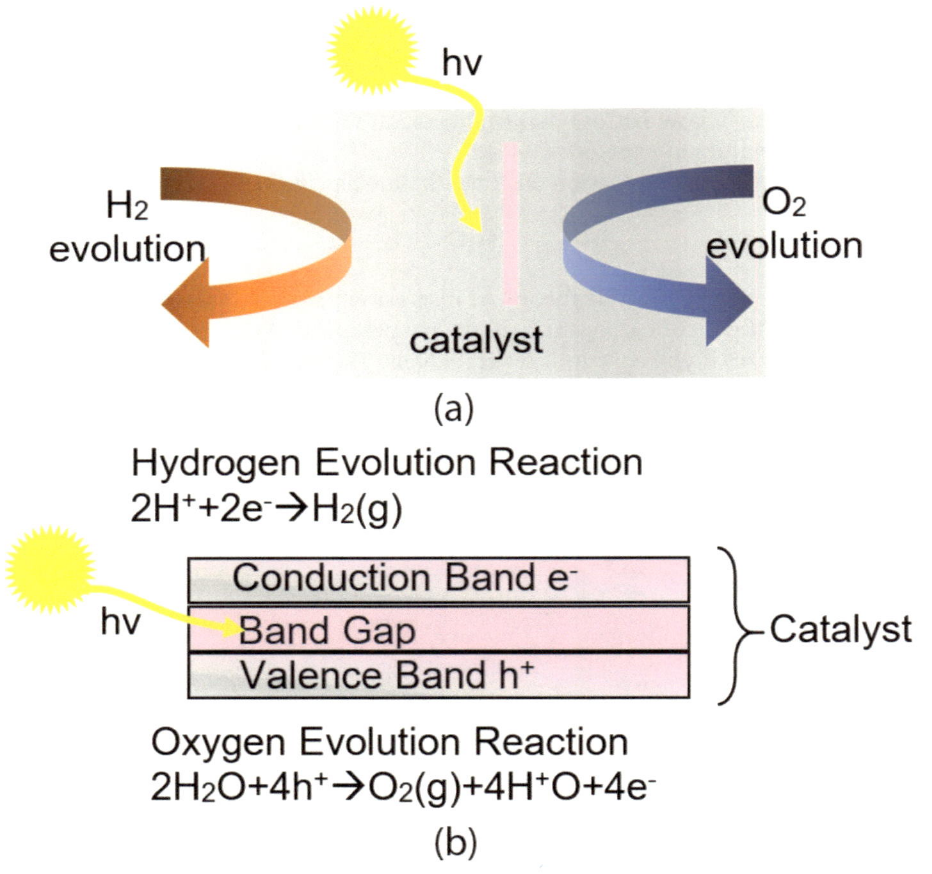

Figure 8.
Schematic of the reaction processes involved in overall water splitting. (a) The processes of photocatalytic water splitting. (b) Water molecules are oxidized by the holes to form O_2 and reduced by the electrons to form H_2.

through compositional variations; perovskites show promise for solar hydrogen production among the large number of photocatalysts being explored. As example of photocatalysts can be mentioned perovskites with formula $AFeO_3$, where A: La, Pr, Ce, and perovskites with formula $LaBO_3$ where B: Co, Mn, Fe [53–55]. Under visible light $LaFeO_3$ with a bandgap of 2.1 eV has an oxygen evolution rate of 331.5 μmol/h*g in ethanol [56]; $SrTiO_3$ with a bandgap of 3.2 eV has a rate of 18.8 μmol/h*g in ethanol [57]. Perovskite materials are also frequently explored in combination with other oxides to carry out various steps of complex uphill reactions involved in water splitting reactions. In **Figure 8**, the schematic of the reaction processes involved in overall water splitting is shown. When the energy of incident light is larger than that of a bandgap, electrons and holes are generated in the conduction and valence bands, respectively.

The perovskites tailored through bandgap engineering approaches can harvest solar light more effectively. Such perovskites show broadband absorption over the visible to near-infrared region of the solar spectrum. Another challenge related to the charge separation between photogenerated holes and electrons has also been undertaken with perovskite compositions to achieve overall improved efficiency.

6. Conclusions and outlook

In the last years environment-friendly and clean energy have attracted worldwide attention due to the growing concerns about global warming and other environmental issues associated with the heavy consumption of fossil fuels. Currently, new functional materials and adaptations to existing functional materials and their use are undergoing extensive investigation and have seen a remarkable development. Perovskites drive interest in their research investigations because the promise of their excellent features to be used in important technological devices such as the solid oxide fuel cell, water electrolysis, and photocatalysts.

We described the observed remarkable feature of adjustable structure properties for perovskites, leaving room for obtaining perovskite oxides with better functional performances. The perovskite structure is viable to wide departures in compositions from the ideal formula ABO_3. The presence of defects can change the properties of the material. When these are properly controlled, defects are the material engineer's way of tuning material properties into wanted effects.

For real world applications, there are still some challenges to tackle in order to achieve the primary goal of the energy field when perovskites with different morphologies are the target materials. Oxide perovskite materials are a competitive alternative to low-cost non-noble metal-based functional materials with high activity and stability to replace the state-of-the-art material. Approaches to prepare perovskites are usually complicated and need harsh reaction conditions like high temperatures, in most of the cases. The large-scale synthesis of perovskites is also a great challenge to researchers since most work is still limited to laboratory scale. Therefore, searching for new facile and environmentally friendly approach to synthesize perovskites is still of great importance in this field.

It can be seen from the above discussion that there are many positive outcomes in the continuous development of perovskite materials rationally engineered by defect chemistry and controlled morphologies through the preparation methods. Great efforts are dedicated for the use of *in-situ* characterization techniques like X-ray diffraction or microscopy, so we can explain how structural modifications, alterations in the morphology, degradation speed of perovskite happened. Moving toward commercialization it is necessary to ensure low cost and long-term stability of materials which represent a significant active research direction.

Acknowledgements

This work was supported by the National Authority for Scientific Research and Innovation/Romanian Ministry of Education and Research, project "RESTORE"—117/16.09.2016 ID/Cod My SMIS: P_37_595 / 104958.

Conflict of interest

The authors state that there is no conflict of interest associated with this work.

Author details

Mirela Dragan[1]*, Stanica Enache[1], Mihai Varlam[1] and Konstantin Petrov[2]

1 National R&D Institute for Cryogenic and Isotope Technologies, Ramnicu Valcea, Romania

2 Academician Evgeni Budevski Institute of Electrochemistry and Energy Systems, Sofia, Bulgaria

*Address all correspondence to: mirela.dragan@icsi.ro

References

[1] Thurnaurer H, Deaderick J. U.S. Patent No. 2,429,588 filed (1941); 1947

[2] Bednorz JG, Müller KA. Possible high T superconductivity in the Ba−La−Cu−O system. Zeitschrift fur Physik B: Condensed Matter. 1986; **64**(2):189-193. DOI: 10.1007/BF01303701

[3] Tonui P, Oseni SO, Sharma G, Yan Q, Mola GT. Perovskites photovoltaic solar cells: An overview of current status. Renewable and Sustainable Energy Reviews. 2018;**91**: 1025-1044. DOI: 10.1016/j.rser.2018.04.069

[4] Alvarez G, Conde-Gallardo A, Montiel H, Zamorano R. About room temperature ferromagnetic behavior in $BaTiO_3$ perovskite. Journal of Magnetism and Magnetic Materials. 2016;**401**:196-199. DOI: 10.1016/j.jmmm.2015.10.031

[5] Zheng T, Wu J, Xiao D, Zhu J. Recent development in lead-free perovskite piezoelectric bulk materials. Progress in Materials Science. 2018;**98**:552-624. DOI: 10.1016/j.pmatsci.2018.06.002

[6] Li X, Jeanloz R. Electrical conductivity of $(Mg,Fe)SiO_3$ perovskite and a perovskite-dominated assemblage at lower. Geophysical Research Letters. 1987;**14**(11):1075-1078. DOI: 10.1029/GL014i011p01075

[7] Presto S, Kumar P, Varma S, Viviani M, Singh P. Electrical conductivity of NiMo-based double perovskites under SOFC anodic conditions. International Journal of Hydrogen Energy. 2018;**43**(9): 4528-4533. DOI: 10.1016/j.ijhydene.2018.01.066

[8] Rubel MHK, Miura A, Takei T, Kumada N, Mozahar Ali N, Nagao M, et al. Superconducting double perovskite bismuth oxide prepared by a low-temperature hydrothermal reaction. Angewandte Chemie International. 2014;**53**(14):3599-3603. DOI: 10.1002/ANIE.201400607

[9] Kim DC, Baranov AN, Kim JS, Kang HR, Kim BJ, Kim YC, et al. High pressure synthesis and superconductivity of $Ba_{1-x}K_xBiO_3$ $(0.35<x<1)$. Physica C: Superconductivity. 2003;**383**(4):343-353. DOI: 10.1016/S0921-4534(02)01332-1

[10] Phraewphiphat T, Iqbal M, Suzuki K, Matsuda Y, Yonemura M, Hirayama M, et al. Syntheses, structures, and ionic conductivities of perovskite-structured lithium–strontium–aluminum/gallium–tantalum-oxides. Journal of Solid State Chemistry. 2015;**225**:431-437. DOI: 10.1016/j.jssc.2015.01.007

[11] Salles C, Bassat JM, Fouletier J, Marinha D, Steil M-C. Oxygen pressure dependence of the ionic conductivity of iron-doped calcium titanate. Solid State Ionics. 2018;**324**:103-108. DOI: 10.1016/j.ssi.2018.06.014

[12] Lin Q, Yang X, Lin J, Guo Z, He Y. The structure and magnetic properties of magnesium-substituted $LaFeO_3$ perovskite negative electrode material by citrate sol-gel. International Journal of Hydrogen Energy. 2018;**43**(28): 12720-12729. DOI: 10.1016/j.ijhydene.2018.03.156

[13] Goswami S, Bhattacharya D. Magnetic transition at ~150 K in nanoscale $BiFeO_3$. Journal of Alloys and Compounds. 2018;**738**:277-282. DOI: 10.1016/j.jallcom.2017.12.107

[14] Zhang F, Zhang X, Jiang G, Li N, Hao Z, Qu S. H_2S selective catalytic oxidation over Ce substituted. Chemical Engineering Journal. 2018;**348**:831-839. DOI: 10.1016/j.cej.2018.05.050

[15] Wang WL, Meng Q, Xue Y, Weng X, Sun P, Wu Z. Lanthanide perovskite catalysts for oxidation of chloroaromatics: Secondary pollution and modifications. Journal of Catalysis. 2018;**366**:213-222. DOI: 10.1016/j.jcat.2018.07.022

[16] Fang M, Yao X, Li W, Li Y, Shui M, Shu J. The investigation of lithium doping perovskite oxide $LiMnO_3$ as possible LIB anode material. Ceramics International. 2018;**44**(7):8223-8231. DOI: 10.1016/j.ceramint.2018.02.002

[17] Yao C, Zhang H, Liu X, Meng J, Zhang X, Meng F, et al. Characterization of layered double perovskite $LaBa_{0.5}Sr_{0.25}Ca_{0.25}Co_2O_{5+\delta}$ as cathode material for intermediate-temperature solid oxide fuel cells. Journal of Solid State Chemistry. 2018;**265**:72-78. DOI: 10.1016/j.jssc.2018.05.028

[18] Parganiha Y, Kaur J, Dubey V, Shrivastavac R. $YAlO_3$:Ce^{3+} powders: Synthesis, characterization, thermoluminescence and optical studies. Superlattices and Microstructures. 2015;**85**:410-417. DOI: 10.1016/j.spmi.2015.06.011

[19] Baig HN, Saluja JK, Haranath D. Investigation of luminescence properties of Dy^{3+} doped $YAlO_3$ phosphors synthesized through solid state method. Optik. 2016;**127**(20): 9178-9195. DOI: 10.1016/j.ijleo.2015.12.159

[20] Zhang Y, Tso CY, Iñigo JS, Liu S, Miyazaki H, Chao CYH, et al. Perovskite thermochromic smart window: Advanced optical properties and low transition temperature. Applied Energy. 2019;**254**:113690. DOI: 10.1016/j.apenergy.2019.113690

[21] Matsouka C, Zaspalis V, Nalbandian L. Perovskites as oxygen carriers in chemical looping reforming process—Preparation of dense perovskite membranes and ionic conductivity measurement. Materials Today: Proceedings. 2018;**5**(14): 27543-27552. DOI: 10.1016/j.matpr.2018.09.074

[22] Baek D, Kamegawa A, Takamura H. Mixed conductivity and electrode properties of Mn-doped Bi–Sr–Fe-based perovskite-type oxides. Solid State Ionics. 2013;**253**:211-216. DOI: 10.1016/j.ssi.2013.09.056

[23] Swierczek K, Zajac W, Klimkowicz A, Zheng K, Malikova N, Dabrowski B. Crystal structure and proton conductivity in highly oxygen-deficient $Ba_{1-x}La_x(In, Zr, Sn)O_{3-\delta}$ perovskites. Solid State Ionics. 2015;**275**: 58-61. DOI: 10.1016/j.ssi.2015.02.018

[24] Zang M, Zhao C, Wang Y, Chen S. A review of recent advances in catalytic combustion of VOCs on perovskite-type catalysts. Journal of Saudi Chemical Society. 2019;**23**(6):645-654. DOI: 10.1016/j.jscs.2019.01.004

[25] De Graef M, McHenry ME. Structure of materials: An introduction to crystallography, diffraction and symmetry. Cambridge, United Kingdom: Cambridge University Press; 2007. p. 671. ISBN: 978-0-521-65151-6

[26] Ali R, Yashima M. Space group and crystal structure of the perovskite $CaTiO_3$ from 296 to 1720 K. Journal of Solid State Chemistry. 2005;**178**(9): 2867-2872. DOI: 10.1016/j.jssc.2005.06.027

[27] Goldschmidt VM. Die gesetze der krystallochemie. Die Naturwissenschaften. 1926;**21**:477-485. DOI: 10.1007/bf01507527

[28] Dragan M, Misture S. In-situ analysis of chemical expansion and stability of SOFC cathodes. Materials Research Society Symposium Proceedings. 2014;**1655**:77-82. DOI: 10.1557/opl.2014.413

[29] Knyazev AV, Mączka M, Krasheninnikova OV, Ptak M, Syrov EV, Trzebiatowska-Gussowska M. High-temperature X-ray diffraction and spectroscopic studies of some Aurivillius phases. Materials Chemistry and Physics. 2018;**204**:8-17. DOI: 10.1016/j.matchemphys.2017.10.022

[30] Zhu C, Nobuta A, Ju YW, Ishihara T, Akiyama T. Solution combustion synthesis of $Ce_{0.6}Mn_{0.3}Fe_{0.1}O_2$ for anode of SOFC using $LaGaO_3$-based oxide electrolyte. International Journal of Hydrogen Energy. 2013;**38**(30):13419-13426. DOI: 10.1016/j.ijhydene.2013.08.007

[31] Fu QX, Tietz F, Lersch P, Stover D. Evaluation of Sr- and Mn-substituted $LaAlO_3$ as potential SOFC anode materials. Solid State Ionics. 2006;**177**(11–12):1059-1069. DOI: 10.1016/j.ssi.2006.02.053

[32] Xie H, Wei Z, Yang Y, Chen H, Ou X, Lin B, et al. New Gd-Zn co-doping enhanced mechanical properties of $BaZrO_3$ proton conductors with high conductivity for IT-SOFCs. Materials Science and Engineering B. 2018;**238–239**:76-82. DOI: 10.1016/j.mseb.2018.12.012

[33] Shimada H, Yamaguchi T, Sumi H, Nomura K, Yamaguchi Y, Fujishiro Y. Extremely fine structured cathode for solid oxide fuel cells using Sr-doped $LaMnO_3$ and Y_2O_3-stabilized ZrO_2 nano-composite powder synthesized by spray pyrolysis. Journal of Power Sources. 2017;**341**:280-284. DOI: 10.1016/j.jpowsour.2016.12.002

[34] Dragan M, Enache S, Varlam M, Petrov K. Perovskite-type material lanthanum cobaltite $LaCoO_3$: aspects of processing route toward practical applications. In: Cobalt Compounds and Applications. Rijeka, Croatia: IntechOpen Limited; 2019. DOI: 10.5772/intechopen.86260. ISBN: 978-1-78984-559-4

[35] Lakshminarayanan N, Kuhn JN, Rykov SA, Millet JMM, Ozkan US. Doped $LaFeO_3$ as SOFC catalysts: Control of oxygen mobility and oxidation activity. Catalysis Today. 2010;**157**(1–4):446-450. DOI: 10.1016/j.cattod.2010.03.037

[36] Miruszewski T, Kamecki B, Lapinski M, Karczewski J. Fabrication, structural and electrical properties of Sr(V,Nb)$O_{3-\delta}$ perovskite materials. Materials Chemistry and Physics. 2018;**212**:446-452. DOI: 10.1016/j.matchemphys.2018.03.070

[37] Yaremchenko AA, Brinkmann B, Janssen R, Frade JR. Electrical conductivity, thermal expansion and stability of Y- and Al-substituted $SrVO_3$ as prospective SOFC anode material. Solid State Ionics. 2013;**247–248**:86-93. DOI: 10.1016/j.ssi.2013.06.002

[38] Jiang SP, Liu L, Ong KP, Wu P, Li J, Pu J. Electrical conductivity and performance of doped $LaCrO_3$ perovskite oxides for solid oxide fuel cells. Journal of Power Sources. 2008;**176**(1):82-89. DOI: 10.1016/j.jpowsour.2007.10.053

[39] Guo P, Zeng C, Shao Y. Effect of $LaCrO_3$ coating on high temperature oxidation of type 316 stainless steel. Journal of Rare Earths. 2011;**29**(7):698-701. DOI: 10.1016/S1002-0721(10)60525-X

[40] Trasatti S. Water electrolysis: Who first? Journal of Electroanalytical Chemistry. 1999;**476**(1):90-91. DOI: 10.1016/S0022-0728(99)00364-2

[41] Enache S, Dragan M, Varlam M, Petrov K. Electronic percolation threshold of self-standing Ag-$LaCoO_3$ porous electrodes for practical applications. Materials. 2019;**12**(15):2359. DOI: 10.3390/ma12152359

[42] Enache S, Dragan M, Soare A, Petrov K, Varlam M. Environmentally

friendly methods for high quality lanthanum cobaltite perovskite catalyst synthesis. Progress of Cryogenics and Isotopes Separation. 2019;**22**(1):39

[43] Enache S, Dragan M, Soare A, Ebrasu DI, Zaulet A, Varlam M, et al. One step solid-state synthesis of lanthanum cobalt oxide perovskites as catalysts for oxygen evolution in alkaline media. Bulgarian Chemical Communications. 2018;**50**(A):127-132

[44] Jiang SP. A comparison of O_2 reduction reactions on porous (La,Sr) MnO_3 and (La,Sr)(Co,Fe)O_3 electrodes. Solid State Ionics. 2002;**146**(1–2):1-22. DOI: 10.1016/S0167-2738(01)00997-3

[45] Manwar NR, Borkar RG, Khobragade R, Rayalu SS, Jain SL, Bansiwal AK, et al. Efficient solar photo-electrochemical hydrogen generation using nanocrystalline $CeFeO_3$ synthesized by a modified microwave assisted method. International Journal of Hydrogen Energy. 2017;**42**(16): 10931-10942. DOI: 10.1016/j.ijhydene.2017.01.227

[46] Petric A, Huang P, Tietz F. Evaluation of La–Sr–Co–Fe–O perovskites for solid oxide fuel cells and gas separation membranes. Solid State Ionics. 2002;**135**(1–4):719-725. DOI: 10.1016/S0167-2738(00)00394-5

[47] Tietz F, Arul Raj I, Zahid M, Mai A, Stover D. Survey of the quasi-ternary system $La_{0.8}Sr_{0.2}MnO_3$–$La_{0.8}Sr_{0.2}CoO_3$–$La_{0.8}Sr_{0.2}FeO_3$. Progress in Solid State Chemistry. 2007;**35**(2–4):539-543. DOI: 10.1016/j.progsolidstchem.2007.01.028

[48] Ullmann H, Trofimenko N, Tietz F, Stover D, Ahmad-Khanlou A. Correlation between thermal expansion and oxide ion transport in mixed conducting perovskite-type oxides for SOFC cathodes. Solid State Ionics. 2000; **138**(1–2):79-90. DOI: 10.1016/S0167-2738(00)00770-0

[49] Choi S, Yoo S, Kim J, Park S, Jun A, Sengodan S, et al. Highly efficient and robust cathode materials for low-temperature solid oxide fuel cells: $PrBa_{0.5}Sr_{0.5}Co_{2-x}Fe_xO_5$. Scientific Reports. 2013;**3**:2426. DOI: 10.1038/srep02426

[50] Dragan M, Enache S, Soare A, Petrov K, Varlam M. $LaCoO_3$ perovskite-type oxide: Synthesis and characterization towards practical applications. Progress of Cryogenics and Isotopes Separation. 2018;**21**(2):49-56

[51] Ebrasu DI, Zaulet A, Enache S, Dragan M, Carcadea E, Varlam M, et al. Electrochemical characterization of metal oxide as catalysts oxygen evolution in alkaline media. Bulgarian Chemical Communications. 2018;**50**(A): 133-138

[52] Humayun M, Sun N, Raziq F, Zhang X, Yan R, Li Z, et al. Synthesis of ZnO/Bi-doped porous $LaFeO_3$ nanocomposites as highly efficient nano-photocatalysts dependent on the enhanced utilization of visible-light-excited electrons. Applied Catalysis B: Environmental. 2018;**231**:23-33. DOI: 10.1016/j.apcatb.2018.02.060

[53] Ibarra-Rodriguez LI, Huerta-Flores AM, Mora-Hernandez JM, Torres-Martínez LM. Photocatalytic evolution of H_2 over visible-light active $LaMO_3$ (M: Co, Mn, Fe) perovskite materials: Roles of oxygenated species in catalytic performance. Journal of Physics and Chemistry of Solids. 2020;**136**:109189. DOI: 10.1016/j.jpcs.2019.109189

[54] Egelund S, Caspersen M, Nikiforov A, Moller P. Manufacturing of a $LaNiO_3$ composite electrode for oxygen evolution in commercial alkaline water electrolysis. International Journal of Hydrogen Energy. 2016;**41**(24): 10152-10160. DOI: 10.1016/j.ijhydene.2016.05.013

[55] Kim J, Chen X, Shih PC, Yang H. Porous perovskite-type lanthanum cobaltite as electrocatalysts toward oxygen evolution reaction. ACS Sustainable Chemistry & Engineering. 2017;**5**(11):10910-10917. DOI: 10.1021/acssuschemeng.7b02815

[56] Tijare SN, Joshi MV, Padole PS, Mangrulkar PA, Rayalu SS, Labhsetwar NK. Photocatalytic hydrogen generation through water splitting on nano-crystalline $LaFeO_3$ perovskite. International Journal of Hydrogen Energy. 2012;**37**(13): 10451-10456. DOI: 10.1016/j.ijhydene.2012.01.120

[57] Puangpetch T, Sreethawong T, Yoshikawa S, Chavadej S. Hydrogen production from photocatalytic water splitting over mesoporous-assembled $SrTiO_3$ nanocrystal-based photocatalysts. Journal of Molecular Catalysis A: Chemical. 2009;**312**(1–2): 97-106. DOI: j.molcata.2009.07.012

Section 3

Other Applications of Perovskite

Chapter 7

Lead-Free Hybrid Perovskite Light-Harvesting Material for QD-LED Application

Rajan Kumar Singh, Neha Jain, Sudipta Som, Somrita Dutta, Jai Singh and Ranveer Kumar

Abstract

Most recently, organic-inorganic semiconductor light harvester materials, have arisen as a new class of functional element and attracts the research community due to its outstanding optoelectronic properties. Organic-inorganic perovskites are solution process that is easy for the fabrication of devices at low temperature. Additionally, up to date, perovskite quantum dots have emerged as the most efficient light harvester for LEDs and display applications, with high color purity, color tunability, and photoluminescence quantum yield up to 100%. However, the presence of lead in organic-inorganic perovskites and the stability issue of perovskite materials are the significant challenges for the research community. To date, some lead substitute materials have been tried to enhance the film morphology and production of the less toxic light harvester. In this chapter, we focus on the lead substitution on B sight with homovalent cations like Sn^{2+}, Mn^{2+}, Cd^{2+}, and Zn^{2+} cations. These lead substitutions not only reduce the toxicity of perovskite material while these dopants also enhance the optical and performance of LEDs. We also included the LEDs application of lead substituted perovskite quantum dots (PQDs) that may be useful for the environmental friendly and highly performing perovskite quantum dot LEDs (PQ-LEDs) shortly.

Keywords: hybrid perovskite, quantum dots, lead-free, light emitting diodes

1. Introduction

Organic-inorganic perovskite light harvesting materials have attracted passionate interest in past few years and raised as one of the most promising solar cells with power conversion efficiency up to 23% as well as excellent performance in LED display devices with high color gamut and color purity [1]. For the photovoltaic (PV) application, perovskite materials were first used in 2009 and for QDs LED applications, perovskite material was used firstly in 2014 [2]. Perovskite-based devices are solution processable that makes it cost-effective and helps in large scale production. Presently, such kind of perovskite solar cells is now comparable to the commercially available solar cells including, silicon, CdTe, CuInGaSe, and GaAs [3]. Perovskite quantum dots (PQDs) also are the most research centered topic in the field of nano-materials and colloidal QDs due to its excellent optical behavior as explained earlier [4].

IntechOpen

After long research and development, semiconductor QDs (CdSe, CdZnSe, InP, etc.) and OLEDs have emerged as the leading technologies in the field of display devices [3]. OLEDs have lit pixels, less energy consuming properties, superior contrast, and wider viewing angles. However, synthesis and fabrication process of OLEDs are very complicated that raise the price of devices. On the other hand, PQDs possess tunable, narrow band emission and high PLQY that significantly improve the performance in both EL and PL devices as explained earlier. Existing CdSe and InP QDs have taken decades to be suitable as the robust candidates for displays. Samsung QLED TVs have been in the forefront serving the premium market along with Chinese brands, TCL and Hisense. Samsung Display is building a pilot production facility for QD OLED, which will start production in 2019. Most research is centered on PQDs due to their high photoluminescence (PLQY), tunable band gaps, and narrow emission wavelength. With quantum confinement effects, the emission of PQDs can be controlled with size and ingredients. Hence, PQDs are useful in photovoltaic cells, laser applications, light-emitting diodes (LEDs), and bio-imaging. Promising color purity and emission tunability of QDs make them promising contenders for next-generation displays as a light converter (white-light LEDs) and active-mode QD-LEDs (QLEDs) [3, 5]. Additionally, PQDs work faster than other QDs due to the absence of a deep state, where the electron-hole pair does not immediately revert to its ground state. In this context, PQDs have become promising candidates for the next-generation of w-LEDs for lighting and other display applications [6–9]. After the discovery of PQDs, more than 7000–8000 peer-reviewed literatures have been published. The fast rate of scientific research publication reveals the interest of the research community in the PQDs topic. Besides excellent performances of PQD technology, some challenges restrict the lead-containing perovskite materials for commercialization. Lead halide perovskite both bulk as well as nanoparticles/QDs are soft and very sensitive with humidity, moisture, light, air, temperature, etc. [10]. Therefore, the most apparent work necessity is to develop the new synthesis routes and new approaches of their more stabilization concerning ambient conditions, water, light, and temperature. Bromide-based perovskite has excellent performance, but the major shortcoming is the stability and presence of toxic element Pb. In this regards, a more in-depth understanding of the mechanism on the radiative and nonradiative recombination is necessarily required [11, 12].

Thus it is an open question that does the organic-inorganic perovskite-based device industry fabrication comes soon? The answer may be controversial because research community will say yes while some will say no because of the toxicity of Pb and long term stability of the materials. According to the U. S. EPA, the maximum amount of Pb^{2+} in air and water should not be more than 0.1 μg L^{-1} and 15 μg L^{-1} respectively [13]. Hence mass production of pure Pb-based perovskite may pollute the Earth due to its toxicity, long degradation lifetime and easy solubility in water. Therefore, we need to develop a new approach to explore environmental friendly Pb free or limited amount of Pb-based perovskite materials for optoelectronic applications. Thus these two main drawbacks of perovskite material restrict it in commercialization. Hence, the demand for nontoxic element-based perovskite materials continues to grow in the past few years [14]. Among these nontoxic alternatives, divalent tin cation has been considered as a right candidate in replacing Pb^{2+}, and the application of tin-halide perovskites in optoelectronic devices has also been investigated. However, Sn^{2+} undergoes facile oxidation to its tetravalent state (Sn^{4+}), creating a high defect density in the perovskite lattice. These defects would generate trap states in the middle of the band gap, leading to rapid non-radiative relaxation of the exciton. The highest PLQY value of tin-based perovskite nanomaterials can only reach 6.4%, which is still far inferior to the lead-based perovskite nanomaterials (~84%) [15]. Potential applications of Sn-based perovskite material in optoelectronic applications, significant efforts have been made recently to improve the photoluminescence properties of tin perovskite-based nanomaterials. Besides Sn, in place of Pb^{2+}, Zn^{2+}, Mn^{2+}, Ge^{2+}, Cu^{2+}, and Bi^{3+} already have been investigated for less lead of lead-free perovskite materials for versatile applications. However, in the case of organic-inorganic halide PQDs, there is very less work reported and some research work based on Pb^{2+} substitution in only inorganic $CsPbX_3$ perovskite systems are going on, and this will be the main focus for the present chapter.

2. Structure of halide perovskite

Perovskites structure derives from the crystal structure of titanium calcium oxide ($CaTiO_3$). This was the first discovered perovskite material in the year 1839 by German mineralogist Gustav Rose and structure of perovskite material was studied by Russian mineralogist Lev A. Perovski [16]. After this discovery, a huge amount of perovskite materials have been found from natural resources and as well as synthesized in the lab also. So perovskite materials can be categorized mainly into two categories, one is based on oxides (oxygen anion) and another based on halides (halogen anions). On the basis of this categorization, perovskite materials adopt ABO_3 and ABX_3 chemical formula, where A and B are the cations and X is halogens such as I, Br, Cl, or mixed halides. ABO_3 structured perovskite is generally used for thermoelectric, superconductive and ferromagnetic applications which were discovered in the 18th century. On the other hand, halide-based perovskite material; $CsPbX_3$ was first reported in 1958, and $CH_3NH_3PbX_3$-based perovskite was seen in 1978. Thus ABX_3 perovskite further can be categorized into two subparts, pure inorganic halide perovskite, and organic-inorganic halide perovskite. Mitzi et al. have first time reported the excitonic property of halide perovskites for light-emitting diodes (LEDs) and thin film transistors in the year 1990 [17]. Nowadays, most extensively studied materials are ABX_3 structured perovskite semiconductor due to its excellent properties. In ABX_3 chemical formula, A, belongs to monovalent inorganic cation such as K^+, Rb^+, Cs^+ or monovalent organic cation like ethylammonium ($CH_3CH_2NH_3^+$, EA), methylammonium ($CH_3NH_3^+$, MA^+), formamidinium ($(CH(NH_2)^{+2})$, FA), B-site stands for divalent cation Pb^{2+}, Sn^{2+}, Ge^{2+}, Ca^{2+}, Sr^{2+}, Ba^{2+} and X is anion (I^-, Br^- and Cl^-). The properties of perovskite material can be easily tuned by changing the A, B or X site in ABX_3 structure [18]. Generally, perovskite materials have cubic structure consists of close-packed AX_3 sub-lattice with divalent B-site cations within the six-fold coordinated cavities. B-X bonding of perovskite governs the electronic behavior of perovskite semiconductors while A cation has no direct role for the electronic properties. But the size of the A cation may cause the distortions in symmetry of the material. Besides the cubic phase, perovskite has tetragonal and orthorhombic phase also. For example, $MAPbI_3$ has a tetragonal structure while $MAPbBr_3$ and $MAPbCl_3$ have a cubic structure at room temperature. The symmetry of $MAPbI_3$ perovskite also depends on the temperature and symmetry increases with increase in temperature such as, at a lower temperature it has octahedral symmetry, and at room temperature to (162.2–327 K) has tetragonal white at a higher temperature above 178.8 K shows cubic phase. $FAPbI_3$ perovskite has higher symmetry than MA-based perovskites hence has been most widely investigated [19]. $FAPbI_3$ perovskite has nearly cubic structure at room temperature and inert atmosphere. At open atmosphere, FA shows its yellow phase that is non-perovskite phase of this material, and it is not useful for optoelectronic devices because of the high band gap.

This is the only drawback with FA material otherwise FA-based perovskite would exceed those of MA [20].

Every ABX_3 structure cannot be a perovskite phase. There is some fixed criteria for the existence of perovskite structure: (1) charge neutrality: that means the charges of cation and anion should be equal. For example, for ABX_3 structure, A and B cations have 1+, and 2+ charges (total = 3+) and X anion has 3− charges. So both cation and anion have equal costs, i.e., N(A) + N(B) = 3 N(X), here N belongs to the valence of the A, B and X ions; (2) the Goldschmidt tolerance factor, t and (3) octahedral factor μ. t and μ are the most potent factor that decides the existence and structure of the perovskite material. Tolerance factor t and octahedral factor μ are defined as the following equation:

$$t = \frac{(rA - rx)}{\sqrt{2}\,(rB - Rx)} \text{ and } \mu = \frac{rB}{rX} \quad (1)$$

where r_A, r_B, and r_X are the ionic radii of A, B, and X, respectively. For the perovskite structure, the value of t and μ must be lies between given limit; $0.89 < t < 1.11$ and $0.442 < \mu < 0.895$ [21]. Thus from the above formula, it is clear that the size of the A, B and X ions play an essential role for the perovskite materials. These days the Goldschmidt tolerance factor concept is beneficial for the finding the lead-free perovskite materials based on the ionic radii of implicated ions. Beside B-site replacement, there are too many possibilities of the other ion positions such as A as well as X-site. On the basis of this concept, Kieslich and co-workers have theoretically found out more than 600 hypothetical perovskites that have not been reported yet [22]. Such kind of study can help develop less lead of lead-free perovskite materials on the practical level for commercialization.

3. Optical properties of halide perovskite

Organic-inorganic perovskite materials demonstrate a strong optical absorption and bandgap tuning, long diffusion length, high charge carrier mobility, ambipolar charge transport and high tolerance of defects. Strong light absorbing properties of the material ideally suited for solar cell devices. On the other hand, band tuning properties and a wide range of light emission in the visible range with narrow emission bandwidth (FWHM) suitable for the light emitting devices. Organic-inorganic perovskite materials have sharp optical band edge that indicates the direct bandgap and minimal disordered materials [23]. The bandgap of perovskite material can be controlled by their chemical compositions and crystallite size. For example, $CH_3NH_3PbI_3$ (red emission), $CH_3NH_3PbBr_3$ (green emission) and $CH_3NH_3PbCl_3$ (blue emission) hybrid perovskites have band gap of 1.55, 2.3 and 3.1 eV, respectively [24].

The band gap energy levels of hybrid perovskite are determined by an anti-bonding hybrid state between the cation B-s and anion X-p orbitals. This state is related to the valance band maxima and conduction band minimum. Thus due to their unique ABX_3 structure and compositional flexibility, optical properties can be tuned by varying elements at each site. The change in I/Br ratio in $CH_3NH_3PbX_3$ perovskite can modify the bandgap from 1.55 to 2.3 eV, i.e., PL emission can be tuned red, orange and yellow. This band gap tuning mechanism is due to the hybridization of p and s orbitals such as Br-4p orbital is overlapped with the I-5p orbital and the Pb-6 s orbital. Similarly, with varying the concentration of Cl/Br, we can tune bandgap from 2.3 to 3.1 eV for hybrid perovskites [25]. Another approach is tuning of the band gap in perovskite structure is substitution of $(CH_3NH_3)^+$ site

(A) by Cs^+, Rb^+ or mixing of both Cs/other alkyl-ammonium cations. With changing the A cation, the bond length with B and X site is also changed that is responsible for the band gap tuning. $CsPbI_3$, $CH_3NH_3PbI_3$ and $FAPbI_3$ has band gap of 1.73, 1.55 and 1.48 eV, respectively [26]. So from these values, we can see that the band gap of the material decreases with increasing the cation size. Besides A and X site, the band gap energy can be tuned with changing the Pb^{2+} (B) from other divalent cations like Sn^{2+}, Mn^{2+}, Cd^{2+}, etc. [26].

Depending on the particle size, dimensions and morphology hybrid perovskite materials show different optical energy gap than their bulk counterparts due to their quantum confinement effect. For traditional nano-materials, quantum confinement effect has been widely studied. In recent times, the development of size-controlled perovskite QDs has enabled exhaustive research and developments of quantum confinement effect in QDs. Size of perovskite QDs can be controlled or tuned by ligand OLA. Friend et al. have studied on the size-dependent photon emission from $CH_3NH_3PbBr_3$ PQDs, where the particle size and the PL emission peak could be tuned by varying the concentration of the perovskite precursors [27]. Quantum confinement effects have also been reported by Kovalenko group in all inorganic $CsPbBr_3$ PQDs. They found that, with a decrease in the diameter of particle size from 11.8 to 3.8 nm, the PL emission peak gradually blue-shifted from 512 to 460 nm [28]. The thickness of the QDs also responsible for the band gap tuning. Thus trough band gap tuning, multi-colored LEDs and white LED can be formed by simple and easy way.

4. One step towards lead-free perovskite materials

Perovskite quantum dots are considered as the most capable aspirant to the next generation of optoelectronic devices and solar cells technology. But the presence of toxicity due to Pb restricts the commercialization. Hence, currently, most of the research is going on less lead or Pb-free perovskites. The nontoxic and environmentally free perovskite can be prepared by substituting Pb with nontoxic elements that can be achieved through two approaches, i.e., homovalent and heterovalent substitution. Homovalent elements for lead substitution with an isovalent cation such as Sn^{2+}, Mn^{2+}, Cd^{2+}, Zn^{2+}, etc. while heterovalent are aliovalent cations like Bi^{3+}, Sb^{3+}. In this work, we will only discuss homovalent cation substitution of PQDs.

4.1 Homovalent substitution

For the Pb-free perovskite materials, many homovalent elements with +2 stable oxidation states can be used. For example, the group- 14 elements Sn^{2+} and Ge^{2+}, i.e., the same group of Pb^{2+} can be the best choice for lead substitution. Beside this, transition metals as Cd^{2+}, Mn^{2+}, Fe^{2+}, Cu^{2+} and Zn^{2+}; alkaline-earth metals such as Ba^{2+}, Sr^{2+} and Ca^{2+}; rare-earth elements like Eu^{2+}, Yb^{2+} can be considered for the Pb-free perovskites [29]. Theoretically, these Pb-substitutes are perfect for perovskite structure according to the tolerance factor calculations. But some elements like Ba, Sr, and Ca have a large band gap, so these are not suitable for semiconducting materials. However, working with Cu and Zn in ambient condition will be difficult. In recent years, Sn^{2+} and Mn^{2+} are the promising candidates for homovalent substitution in perovskite quantum dots.

4.2 Sn-based perovskite quantum dots

The most suitable Pb-substitute is Sn^{2+} for lead-free perovskite because of the around same ionic radii, binding energy and the same electronic configuration of

s^2 valence [29]. $MASnI_3$ is the most studied lead-free perovskite material. Sn-based perovskites are direct bandgap semiconductors, like Pb-based perovskites. This means, the valence band maximum and conduction band minimum recline at the same position in k-space (reciprocal space). Similar to the $APbX_3$ structure, the structural, optical and electrical properties of $ASnX_3$ perovskite is also affected by the size of A-site cation and X-site anion. $MASnI_3$ perovskite also has a tetragonal structure, but the band gap is near unity that is lower than those of the $MAPbI_3$ [30]. Due to the quantum confinement effect, PQDs shows different behavior from their bulk counterpart.

In case of the pure inorganic perovskite $CsPbI_3$, band gap of PQDs is 1.85–1.97 eV. Such kind of higher bandgap highly encouraged to find some lower Pb-containing, low band gap, and more stable perovskite material. $CsSnI_3$ pure inorganic can be the satisfied the all requirements for the good PQDs because of its lower binding energy and broad absorption from visible to near infrared region. Due to this different behavior, such kind of perovskite can exhibit higher light harvesting efficiency than other perovskite materials [31]. However, the stability of the Sn^{2+}-based perovskite material is the biggest issue due to higher sensitivity to air, moisture and sometimes with nonpolar organic solvents also [32]. In recent years, many significant approaches have been tried for the stability of the Sn^{2+} perovskites. For example, $SnBr_2$, $SnCl_2$, and SnF_2 had used as a dopant into the $CsSnI_3$ perovskite for the stability of the materials. In all, depends, SnF_2 and $SnCl_2$ were found more suitable for the air resistive property. However, stability of the Sn-based material was not satisfactory due to the degradation of the sample within only 3hs [33–35].

In the year 2017, Liu et al. reported, the alloyed $CsSn_{1-x}Pb_xI_3$ PQDs, which was more phase stable than its parent $CsPbI_3$ and $CsSnI_3$ PQDs for months in ambient air [36]. They have prepared $CsSn_{1-x}Pb_xI_3$ PQDs via the standard Schlenk line method under a nitrogen atmosphere. In this method, SnI_2/PbI_2 mixture was rapidly dissolved in trioctylphoshine into the cesium olate at 120–170°C and purified by antisolvent methyl acetate (MeOAc) washing process. The 140°C reaction temperature was found the best for homogeneous and proper morphology of $CsSn_{1-x}Pb_xI_3$ PQDs. With the incorporation of Sn, red shifting occurred that can be observed in absorption and PL spectra as shown in **Figure 1a**.

Additionally, the increase in Sn concentration drastically enhanced the light absorption of $CsSn_{1-x}Pb_xI_3$ PQDs; however, PLQY was found up to 3% only. On the other hand, $CsPbI_3$-based PQDs had up to 100% efficiency to date. Low PLQY with Sn-based PQDs indicates the increase in intrinsic defects that were allied with Sn vacancies. Such kind of vacancies creates deep level defects which act as non-radiative recombination centers and hence this reduces the efficiency of PQDs. **Figure 1b** also suggested the average lifetime of $CsSn_{1-x}Pb_xI_3$ PQDs dramatically decreases with increasing the concentration of Sn. This behavior of Sn containing materials also confirms the occurrence of a large density of quenching defects that support the lower PLQY of the PQDs. Liu also reported that $CsSn_{0.60}Pb_{0.40}I_3$ PQDs was best for performance and stability because of the less- Pb content and widest range of light absorption among the alloyed PQDs. Phase, as well as chemical Stability of $CsSn_{0.60}Pb_{0.40}I_3$ PQDs, is confirmed via XRD and XPS analysis (see **Figure 1c** and **d**). XRD **Figure 3b** suggested that there is no degradation was absorbed up to 5 months that is the first time reported such long time stability with Sn-based material.

Furthermore, there was not found Sn^{4+} oxidation stage in the XPS analysis of the sample. In **Figure 1d** XPS graph showed 2 spectra at 486.2 and 494.5 eV that is associated with Sn^{2+} $3d_{5/2}$ and $3d_{3/2}$ state, while deficiency of spectra at 486.9 indicates the absence of Sn^{4+} in perovskite compound. Thus this less lead-containing and air-stable Sn-based PQDs can be the best choice for solar cell devices but for highly

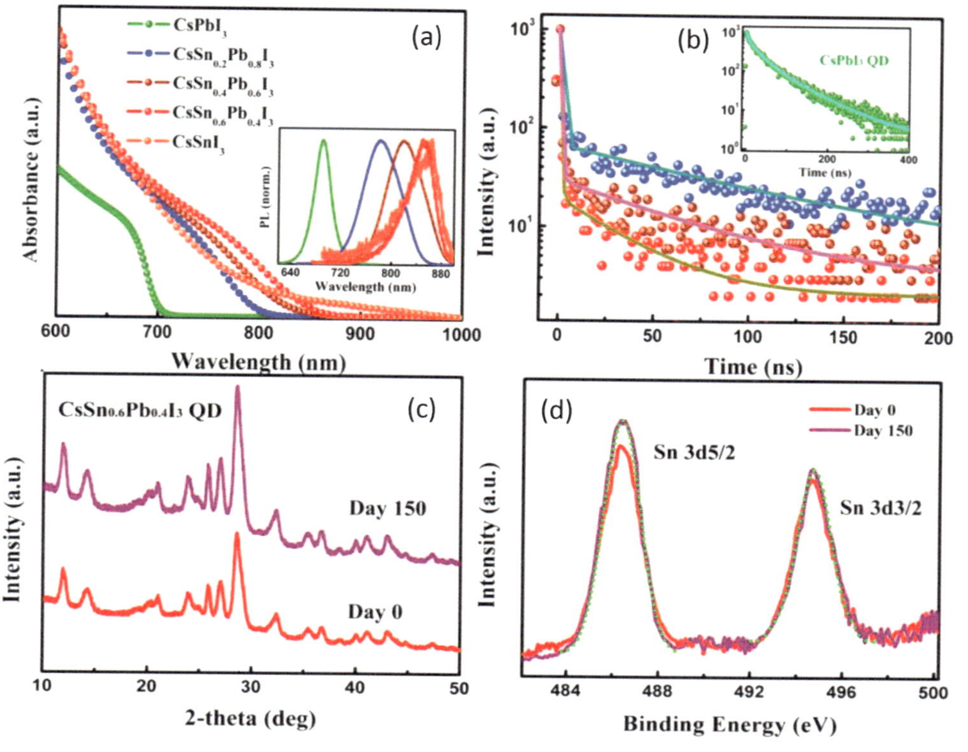

Figure 1.
(a) Optical absorption spectra of different Sn doped PQDs and the inset fig shows their corresponding normalized steady photoluminescence spectra, (b) time-resolved photoluminescence decay for various stoichiometries of Sn, (c) X-ray diffraction patterns and (d) XPS spectra for the $CsSn_{0.60}Pb_{0.40}I_3$ *PQDs (as prepared) and those after storage for 150 days.*

efficient LED there will be a need in the improvement of its PLQY. Jellicoe and co-workers had also tried for the Sn-based perovskite, $CsSnX_3$ (X = Cl, Br, I) for the lead-free perovskite PQDs but due to the instability of Sn it cannot be used for practical applications [35]. To overcome the stability problem, Wang group had replaced Sn^{2+} with Sn^{4+} and developed stable $CsSnI_6$ PQDs with cubic structure. However it's PLQY was just 0.48% that is also not sufficient for optoelectronic applications [36]. From the several studied we absorbed that, $CsSnI_3$-based perovskites have low PLQY but good hole conducting properties. Hence it is mostly used for solar cell applications. For sufficient PLQY and more stable Sn-based material was still needed for display devices. For this purpose, $CsSnBr_3$-based perovskite was found more suitable candidate than iodine-based perovskites.

Yang group have reported the fast and less Pb or Pb- free perovskite nanocrystals by introducing the Sn in Pb site and I at Br site [37]. They have used one pot hot injection method for preparation of SnI_2 doped $CsPbBr_3$ perovskite nanocrystals. After cation and anion exchange between the $CsPbBr_3$ and SnI_2, the lead-free $CsSnI_3$ NCs were obtained. These Pb-free NCs maintained the high PLQY as well as morphology in an inert atmosphere. **Figure 2a** indicated that, with increasing the amount of SnI_2 into $CsPbBr_3$ colloidal solution, dramatically red shifting observed in PL emission and absorption spectra. There was one fascinating result was reported regarding the stability of Sn^{2+}. XRD result predicted that Sn^{2+} converted into Sn^{4+} and formed the small amount of Cs_2SnI_6 phase that indicated the cation as well as anion exchange in the sample. After cation exchange between Pb^{2+} and Sn^{2+} and anion exchange between Br^{-1} and I^{-1}, the PLQY of $CsSnI_3$ and $CsPbBr_3$ NCs was found 59.1 and 73.47% respectively. From this result, the author concluded that the ion exchange did not cause the structural and surface defects. **Figure 4b** shows the schematic representation of the cation and anion exchange in $CsPb_xSn_{1-x}(Br_yI_{1-y})_3$ NCs. Halide anions have high migration and the sufficient vacancy diffusion ability that is why after incorporation of SnI_2 into $CsPbI_3$, halide anions reacts very fast [38]. On the other hand, cation takes more time to reaction with parent material due to the obligation of high activation energy. Moreover, in most of the cases, cation substitution is obsessed with the concentration of halide vacancies [39]. Thus with this cation and anion approach creates a possibility of higher PLQY of Pb-free perovskite NCs.

The 100% Pb replacement with any other divalent cation is impossible till date because of the lower conductivity of substitute element like Sn. In the case of the Sn lower ionic conductivity of PQDs causes many surface defects that make material unstable. Some of the Pb-free perovskite with Sn^{4+} support that instability and toxicity problem can be resolve by partial substitution of Pb^{2+} with Sn^{4+}. In this regards, Liu et al. have reported the less lead-containing $CsPb_{1-x}Sn_xBr_3$ PQDs with PLQY up to 83% [40]. They have performed cation exchange via replacing Pb^{2+} cation with Sn^{4+} cation using hot injection method. Partial replacement with Sn^{4+} in $CsPb_{1-x}Sn_xBr_3$ exhibited the enhancement in PL emission and after some time increasing the more concentration of Sn^{4+} ion, decreases the PL performance due to the increasing the impurity phase of Cs_2SnBr_6. In this work, they found $CsPb_{0.67}Sn_{0.33}Br_3$ was the best composition, and PQDs of material, displayed the high external quantum efficiency and current efficiencies. **Figure 3a** Shows the variation of PL spectra of $CsPb_{1-x}Sn_xBr_3$ PQDs and digital picture in UV irradiation (365 nm). This PL spectrum show the highest PL intensity for x = 33% while increasing the concentration of Sn; PL intensity starts to decrease due to impurity and more surface defects. **Figure 3b** shows the absorption and PL emission spectra of samples that indicated the absence of blue shift that means there is no big effect on band gap after Sn^{4+} incorporation, but exciton recombination can affect the

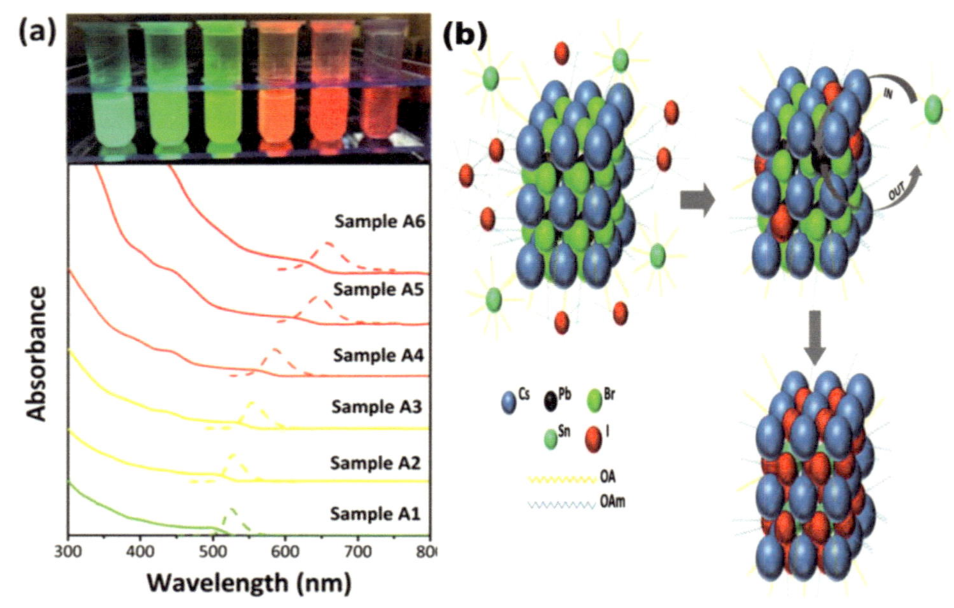

Figure 2.
(a) UV-Visible and PL spectra of different Sn doped $CsPbBr_3$ NCs (A1–6); the colored image of PQDs under 365 nm UV radiation and (b) schematic diagram of the ion exchange process in SnI_2 doped $CsPbBr_3$ NCs.

lifetime and lower PLQY with a higher concentration of Sn^{4+}. A higher concentration of Sn^{4+} also affected the size; morphology and structure of the QDs (see **Figure 3c–g**). After x = 0.33 Sn^{4+} amount in $CsPbBr_3$, the impure phase can be seen in the XRD spectrum and irregular shape and agglomerated particles in TEM due to Cs_2SnBr_6 byproduct.

4.3 Mn^{2+}-based perovskite QDs

Mn^{2+} is a transition metal cation, and doping of Mn^{2+} in II-VI semiconductors has been widely investigated due to its excellent optical, electrical and magnetic properties. All ready many studied have been done on Mn^{2+} doped ZnS, CdSe, CdS, and ZnSe inorganic QDs in which Mn^{2+} incorporation can enhance the long-lifetime and interaction between host d-electron of Mn dopants [41]. In compare to the traditional II-VI group semiconductors, the soaring tolerance in perovskite can be the better candidate for promoting the exciton energy transfer to Mn d-d emission to Mn d-state.

The band gap of Mn^{2+} doped in perovskite materials depends upon the presence of halide anions in a host as well as a dopant. For example, change from Cl^- to Br^- to I^-, the PL emission of perovskite is tuned from blue region to red. In the year 2018, Zhao et al. have reported the Mn^{2+} doped $CsPbCl_3$ NCs and after that intensive exploration are carried on this work [42]. After incorporation of Mn^{2+}

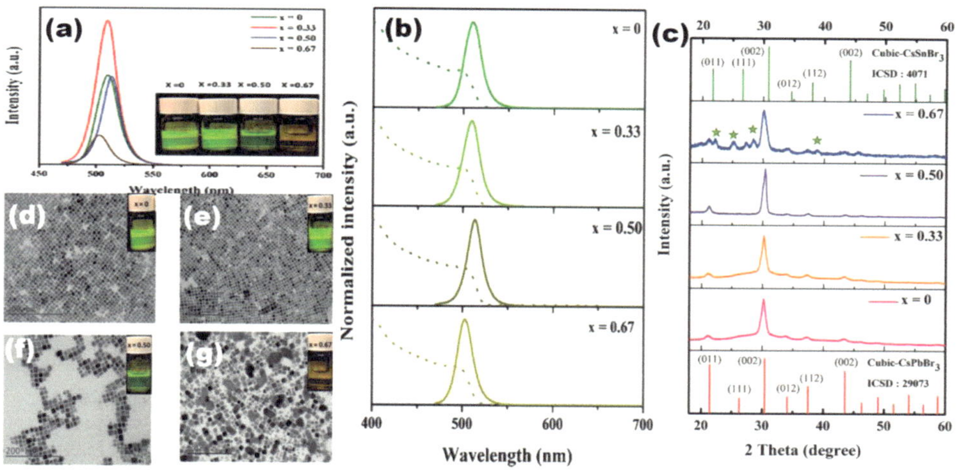

Figure 3.
(a) Variation in PL intensity of different Sn^{4+} doped $CsPbBr_3$ PQDs. Inset of samples under 365 nm UV light (b) UV-Visible absorbance and PL spectra of $CsPb_{1-x}Sn_xBr_3$ (c) XRD pattern of $CsPb_{1-x}Sn_xBr_3$ PQDs, impurity peaks are shown by stare sign. (d–g) TEM images with (d) x = 0, (e) x = 0.33, (f) x = 0.50 and (g) x = 0.67.

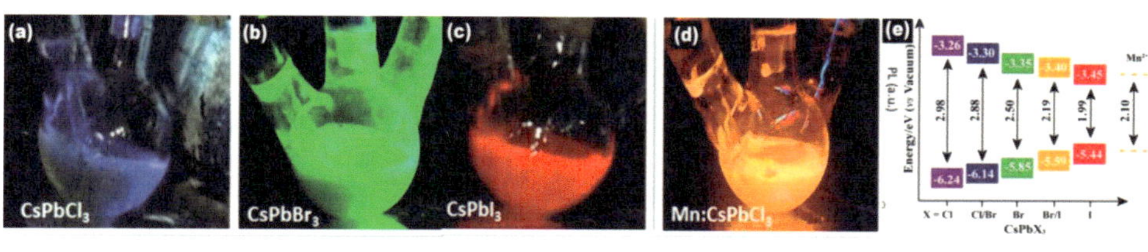

Figure 4.
(a–d) Digital images of $CsPbCl_3$, $CsPbBr_3$, $CsPbI_3$ and Mn: $CsPbCl_3$ nanocrystals under 365 nm UV radiation and (e) band positions of $CsPbX_3$ and Mn^{2+} d-state.

cation in place of Pn^{2+} site, the XRD spectra shied towards higher angle due to the smaller radii of Mn^{2+} while PL emission shifted towards higher wavelength because of the d-d transition. Hence undoped $CsPbCl_3$ NCs shows blue emission, but after doping of Mn^{2+}, it tuned into orange emission as shown in **Figure 4a–c**. It was reported that Cl-based Mn dopant and perovskite host is the best rather than other manganese (II) salts [43]. Via hot injection method, it is tough to doping of Mn^{2+} into $CsPbBr_3$ and $CsPbI_3$ perovskite structure but through the anion exchange process Mn doping is possible. Such kind of possibility depends upon the bond strength and dissociation energy between Mn-X and Pb-X bond. **Figure 4d** shows the band gap energy diagram for Mn-doped different perovskite host [44]. As we know Cl-based perovskite NCs has very low PLQY (<5%) in compared to Br- and I-based perovskites. Besides reduction in toxicity, Mn^{2+} doping helps to enhance the PLQY and lifetime of the perovskite NCs. In this regards, Liu and coworkers fabricated $CsPb_xMn_{1-x}Cl_3$ PQDs via hot injection method [45]. Mn^{2+} incorporation in $CsPbCl_3$ not only increases the PLQY of host material from 5 to 54% but also created an additional intense PL emission peak at 580 nm that is stand for bright orange color (see **Figure 5a**). With increasing the Mn^{2+} concentration secondary, PL peak gradually shifted from 569 to 587 nm. The highest PLQY (54%) was found for the 46% Mn-doped $CsPbCl_3$ perovskite. The various Mn^{2+} compositions with tunable emission are shown in schematic **Figure 5b**. $CsPb_xMn_{1-x}Cl_3$ PQDs has two emission peaks with two different lifetime values such as 13.9 ns at 390 nm and 1.6 ms at 580 nm as shown in **Figure 5a** and **b**. The long decay time for second emission peak near 580 nm is assigned to emission from d-d transition energy transfer mechanism for Mn^{2+} doped semiconductors. Hence Mn^{2+} doping also increases

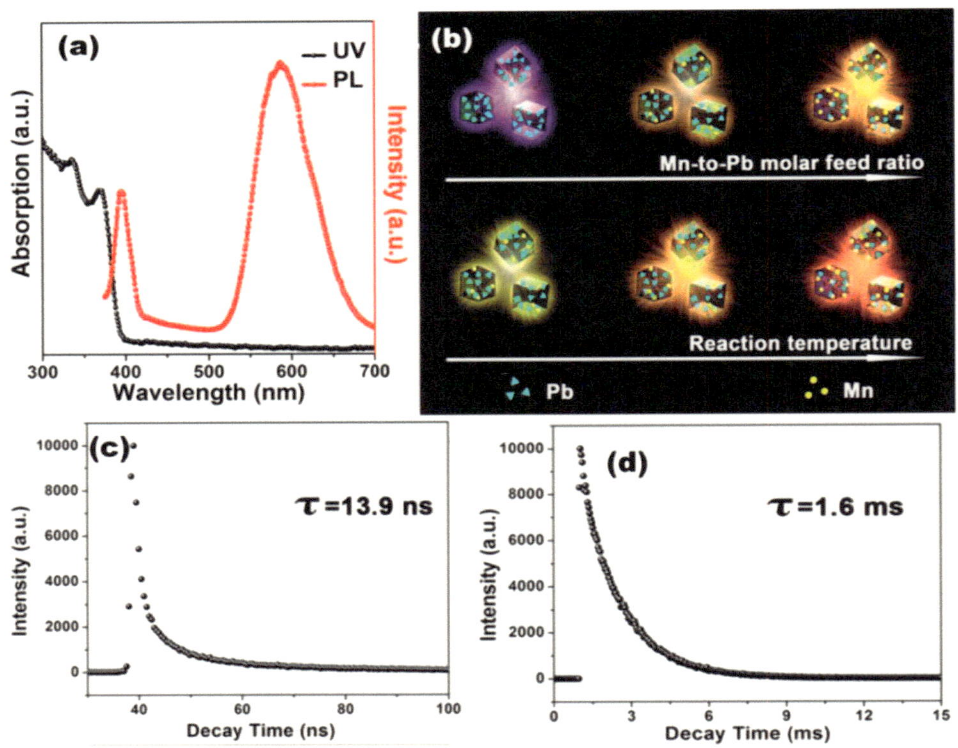

Figure 5.
(a) UV-Visible and PL spectra of $CsPb_{0.54}Mn_{0.46}Cl_3$ PQDs (b) schematic illustration of Mn incorporation in $CsPbCl_3$ PQDs by different molar ratio and reaction temperature, (c and d) PL lifetime of $CsPbCl_3$ and $CsPb_{0.73}Mn_{0.27}Cl_3$ PQDs.

the overall lifetime of PQDs that may be an improvement in the stability of PQDs also. A similar study was carried by Yuan et al. in which they have obtained PLQY of up to 60% for 14.6% Mn^{2+} doped $CsPbCl_3$ [46]. With the help of ligand assisted room temperature method (LART), same work was reported by Zhu group with maximum 41.6% PLQY for 28% Mn^{2+} incorporation in place of Pb^{2+} [47]. Through LART, Mn^{2+}:$CsPbCl_3$ can be prepared in concise time duration and room temperature that reduce the cost of experiment, easy, simple method and high production of products.

In some cases, Mn^{2+} is beneficial for stability of $CsPbX_3$ PQDs. Zou et al. have found that Mn^{2+} incorporation in place of Pb^{2+} not only reduced toxicity while also enhance the thermal as well as air stability of the perovskite samples [48]. Similar to previous work, Zou group also have worked with $CsPbCl_3$: Mn^{2+} PQDs and studied different structural and optical enhanced properties. Besides this work, they have also provided experimental data for $CsPbBr_3$: Mn and $CsPbI_3$: Mn PQDs to prove the absence of energy transfer in Br and I containing perovskites. **Figure 6(a–c)** show the comparison of PL spectra for $CsPbCl_3$: Mn, $CsPbBr_3$: Mn and $CsPbI_3$:Mn PQDs with various Mn^{2+} concentrations. In this case, PL emission intensity first increases with increasing the Mn^{2+} concentration and then decreases with further increasing doping concentration which suggests that the Mn^{2+} ions have a noteworthy impact on the optical properties of PQDs. Presence of additional peak in $CsPbCl_3$: Mn, is the proof of energy transformation, on the other hand, there is no peak shift in Br and I containing PQDs samples due to the mismatch between their optical bang gap absorption and $^4T_1 \rightarrow {}^6T_1$ transition of Mn^{2+} [48]. PLQYs of the Cl-based increases and in Br and I increases up to 20% Mn^{2+} doping and after that dramatically decreases with increasing the Mn^{2+} concentration (see **Figure 6d–f**) while lifetime of $CsPbBr_3$: Mn^{2+} continuously decreases as shown in **Figure 6g**. Mn^{2+} doping in $CsPbBr_3$ and $CsPbI_3$ PQDs also enhance air stability as demonstrated in **Figure 6h** and **i**. **Figure 6h** indicates, the degradation of pure $CsPbBr_3$ PQDs after 30 days while Mn^{2+} doped PQDs is stable up to 120 days while 50%Mn: $CsPbI_3$ PQDs was found stable up to 4 days. As we know $CsPbI_3$ PQDs are unstable in the air, so it is challenging of use for optoelectronic devices. Enhanced stability of Mn: $CsPbI_3$ materials may open the new door for stable red PQDs. The same methodology was reported by Manna group by theoretical and experimental approach also [49]. They reported that unstable $CsPbI_3$ perovskite could be stabilized by incorporation the 10% of Mn^{2+} and there were no major changes in structural and optical properties.

4.4 Cd^{2+} and Zn^{2+} doped perovskite quantum dots

Cd and Zn^{2+} Cation exchange in traditional NCs have been extensively studied, but for perovskite material, there are few reports are present till date. A detailed study of cation exchange has not yet been explained, but halide exchange has been explained well. Halide exchange is easier than cation exchange because of low activation energy and diffusion of anion vacancies in the perovskite materials. Stam et al. have experimentally proved that cation exchange in perovskite takes long time [50]. They reported the cation exchange such as Sn^{2+}, Cd^{2+}, and Zn^{2+} in place of Pb^{2+} in $CsPbBr_3$ perovskite host. 10% cation exchange in perovskite system results in the reduction in toxicity due to lead and also maintained the good PLQY as well as high color purity. But with these, cation exchange, the blue shift was observed in PL emission due to smaller ionic radii of Cd and Zn than Pb (as shown in **Figure 7k**). On the other hand, there is very less shift was obtained in Sn^{2+} due to similar ionic radii as Pb. The Cd^{2+} doping in $CsPbBr_3$ PQDs was resulting in PL emission variation between 452 and 512 nm while Zn showed between 462 and 512 nm. Furthermore, over 60% of PLQYs was obtained for cation doped PQDs and good

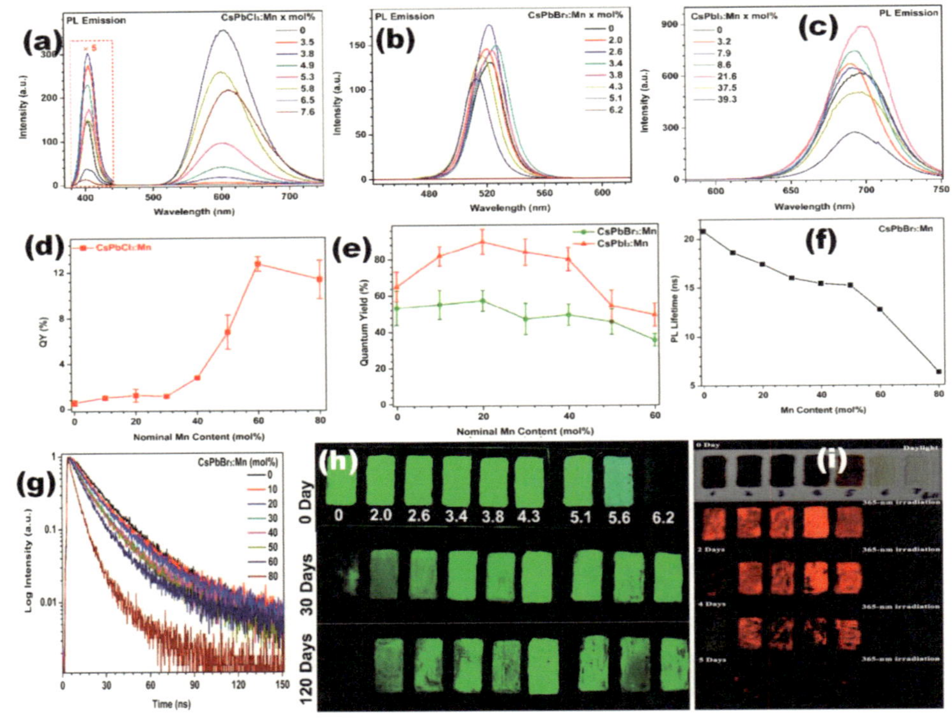

Figure 6.

Comparison of photoluminescence (PL) emission spectra for (a) $CsPbCl_3$:Mn, (b) $CsPbBr_3$:Mn and (c) $CsPbI_3$:Mn QDs doped with different Mn^{2+} contents (d) absolute PL quantum yields (QYs) for $CsPbCl_3$:Mn QDs doped with different nominal Mn^{2+} contents ranging from 0 to 80 mol% (e) absolute PL QYs for $CsPbBr_3$:Mn and $CsPbI_3$:Mn QDs doped with different nominal Mn^{2+} contents ranging from 0 to 60 Mol% (f and g) PL decay curves (left) and lifetimes (right) for excitonic luminescence of $CsPbBr_3$:Mn QDs upon excitation by a 397-nm pulsed laser (h) PL emission photographs for $CsPbBr_3$:Mn QDs coated on the surface of a glass slide with different Mn^{2+} contents from 0 to 6.2 mol% taken under UV irradiation at indicated time periods and (i) red PL emission photographs for $CsPbI_3$:Mn QDs with different nominal Mn^{2+} doping concentrations of 0, 20, 40, 50, 60, 80, 100 mol% from left to right respectively, taken at daylight or under UV irradiation at indicated time period.

stability in ambient conditions. Cation exchange in PQDs reduced the PL lifetime of the $CsPbBr_3$ PQDs. **Figure 7a–j** TEM and energy dispersive X-ray spectroscopy mapping of Cd^{2+} and Zn^{2+} doped $CsPbBr_3$ PQDs. EDS mapping helped to identify the presence of different doped elements in PQDs. **Figure 7a–e** shows the Cd and **Figure 7f**-jshows the Zn doped $CsPbBr_3$ PQDs. These mapping analyses also predicted the uniform distribution and lower concentration of dopant into the host material. Zn^{2+} incorporation enhanced the stability of $CsPbI_3$ black phase in the air also due to improvement in lattice contraction and alloy form of perovskite.

As it is well reported that lower PLQYs of Cl-based perovskite materials due to the large band gap. There are many have published on Mn^{2+} doped $CsPbCl_3$ PQDs, but with doping of, we get orange color so in this case to obtain pure blue color is impossible. On the other hand, other heterojunction-based PQDs such as $CH_3NH_3Bi_2X_9$, $Cs3Bi_2Br_9$ and $Cs_3Sb_2Br_9$ have been developed for blue emission [50]. But still, these blue color emitted PQDs have 46–52% and very unstable that is not suitable for blue LEDs fabrication. To overcome this issue, Mondal et al. reported the effect of Cd incorporation in $CsPbCl_3$ PQDs by hot injection method to enhance the stability and PLQYs of blue color PQDs [51]. For $CdCl_2$ treatment, $CsPbCl_3$ colloidal solution with $CHCl_3$ was mixed with $CdCl_2$ solution ($CdCl_2$ solution was prepared in ethanol) and sonicated for 2–3 min. The obtained colloidal solution was

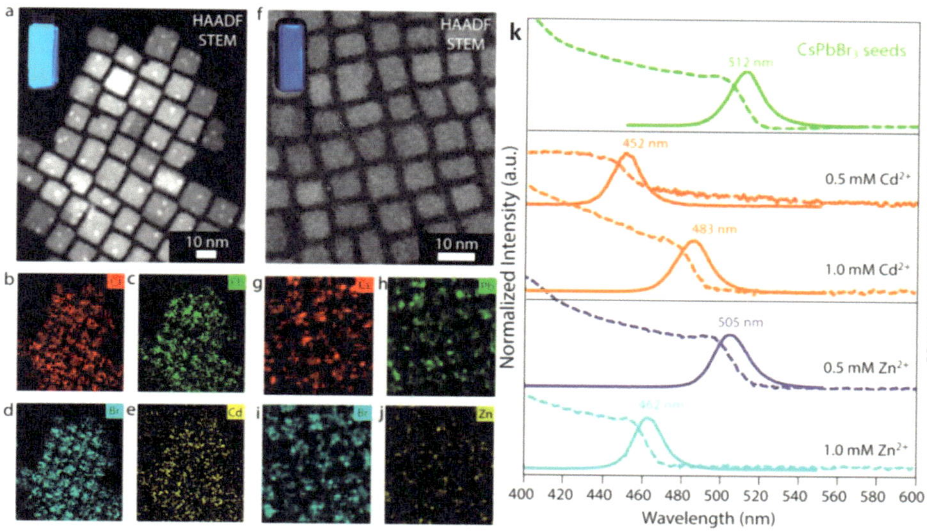

Figure 7.
Energy dispersive X-ray spectroscopy mapping of $CsPb_{1-x}Cd_xBr_3$ and $CsPb_{1-x}Zn_xBr_3$ nanocrystals. (a) HAADF-STEM image of $CsPb_{1-x}Cd_xBr_3$ NCs and corresponding maps of (b) Cs, (c) Pb, (d) Br, and (e) Cd, demonstrating the presence of Cd in the perovskite NCs. The inset in panel shows a photograph of a colloidal suspension of the NCs under UV illumination. (f) HAADF-STEM image of $CsPb_{1-x}ZnxBr_3$ NCs and the corresponding maps of (g) Cs, (h) Pb, (i) Br, and (j) Zn, indicating the presence of Zn in the perovskite NCs. The inset in panel f shows a photograph of a colloidal suspension of the NCs under UV illumination (k) parent $CsPbBr_3$ NCs (green lines) and product NCs obtained after reaction with different concentrations of $CdBr_2$ (orange lines) and $ZnBr_2$ (blue lines).

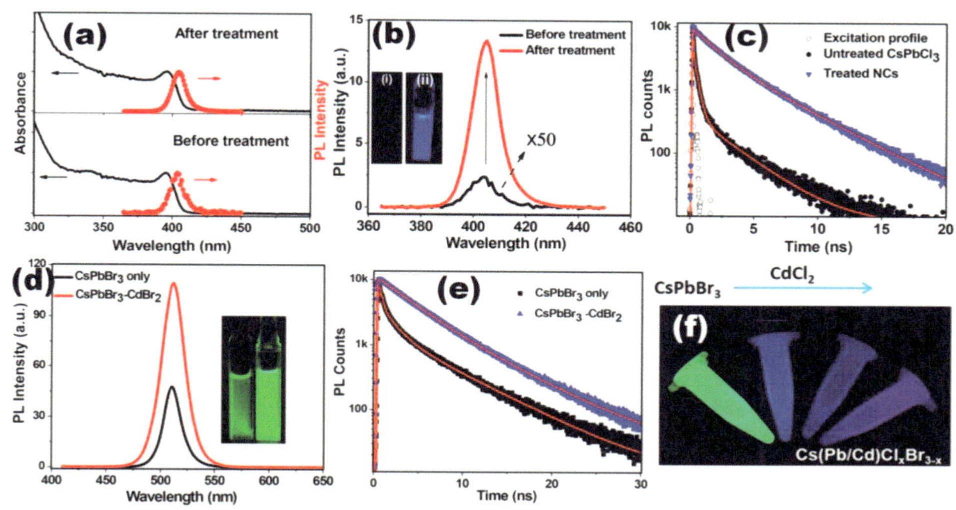

Figure 8.
UV-Visible absorption and PL spectra (a) of the $CsPbCl_3$ NCs before and after the $CdCl_2$ treatment. PL spectra showing dramatic enhancement of PL upon treatment (b) the inset shows the untreated (i) and treated samples (ii) under a UV lamp. (c) PL decay dynamics (λ_{ex} = 375 nm, λ_{PL} = 406 nm) of $CsPbCl_3$ NCs before and after the treatment with $CdCl_2$. PL spectra (d) and PL dynamics (e) of $CsPbBr_3$ NCs before and after the CdBr2 treatment. (f) Change in PL on addition of $CdCl_2$, starting from treated $CsPbBr_3$ NCs (left) to increased addition of $CdCl_2$ (right).

purified by centrifugation with methyl acetate treatments. The $CdCl_2$ treatment of $CsPbCl_2$ dramatically enhanced the from 3–96% PLQYs and narrow PL emission spectra that also indicates the higher color purity of the treated sample.

There are the best thing is that there was no PL or absorbance spectrum shifting observed with $CdCl_2$ treated samples as shown in **Figure 8a**. From **Figure 8b** also indicates the huge improvement in PL emission and bright blue emission under UV light radiation compared to untreated $CsPbCl_3$ PQDs. Generally, with doping of Cd^{2+} or other lower atomic radii elements than Pb^{2+}, has PL emission shift towards blue region due to lower ionic radii. But in this work, the authors did not get any shifting. $CdCl_2$ incorporation not only enhances the PLQYs while it also improved the lifetime and stability of $CsPbCl_3$ PQDs by the four times of the untreated PQDs (**Figure 8c**). A similar strategy was applied with $CsPbBr_3$ PQDs and improvement in PLQY and lifetime also absorbed in green PQDs also (**Figure 8d–f**). Thus this work suggests the Pb^{2+} cation exchange with Cd^{2+} not only provide high PLQYs while long term stability without disturbing any peak shifting. So such kind of blue and green PQDs can be beneficial for good quality of blue, green or white LEDs and a backlight system. Thus Cd and Zn doped PQDs may be useful for different optoelectronic applications similar to pure PQDs due to its low toxicity and excellent properties.

5. Applications of lead substituted PQDs for LEDs

Lead halide-based perovskite QDs, as a promising light-harvesting material for light absorbing and converting light energy, have attracted research community as well as industrialists due to unique properties of PQDs for solid-state lighting (SSL) and flexible color tuning thin film display application. In recent years, organic LED (OLED) and quantum dot LED (QDLED) join the competition of display market and mean to wrest the dominance from LCD. OLED has the advantage of self-luminous, large area fabrication, fast response time, high contrast and application on flexible substrates [2–4]. Due to such type of benefits, many manufacturers have invested in the development of OLEDs, and now various products based on OLEDs have been commercially available. However, there are some limitations with OLEDs like wide PL emission wavelength, instability of organic molecules, etc. Due to this issue, it is difficult to achieve high color purity and a high-quality full-color display. So it is still required to develop a new variety of display technology with high color purity and PLQY to meet the higher demand of consumers. To overcome the less color purity problem, quantum dots can be used because of its excellent luminescence and color characteristics as we have already discussed in the previous section. There is a chance to achieve the high-resolution color contrast and a better full-color display device in near future moreover in contrast to another semiconductor (inorganic) employed in solid state lighting device perovskite material have also benefited regarding the synthesis and manifesting process. PQDs can also fabricate at low temp, solution processed approach to precisely control the size, shape, and purity of perovskite material which makes PQDs ideal candidates for the display device. In addition, due to the colloidal solution characteristics, PQDs is cheaper to use, easier to process and easier to fabricate in large area, large scale production thus PQDs are often touted as disruptive material that could completely replace traditional inorganic phosphor LEDs or OLEDs, QLEDs can be mainly divided into types one is photoluminescence QD LEDs (based on photoexcited), and another is electroluminescence (EL) QDLEDs (electron- excited QDs) [3]. PL QDLEDs is the most commonly used a type of LEDs. QDs LEDs usually have applications in the high-quality display, high resolution, and high contrast display device as well as in lighting application. For commercialization, much less lead-containing perovskite QDLEDs have been investigated in recent years. Lead substituted perovskites also used for PL LEDs and electroluminescence LEDs.

Zou et al. fabricated the EL-based LEDs using $CsPbBr_3$: Mn perovskite QDs [48]. Active LEDs was fabricated by spin coating of PEDOT: PSS on ITO glass substrate as a hole transporting layer, Mn-doped $CsPbCl_3$ as the active layer and TPBi, LiF/Al were used as electrodes by thermal evaporation technique. The schematic illustration of the device is shown in **Figure 9a**. This active led produced electroluminescence with green color (512–515 nm) for $CsPbBr_3$ and Mn-doped perovskites with high color purity and narrow emission spectrum FWHM of 20 nm (see **Figure 9b**). There were no significant shifting observed in EL as well as in current density-voltage properties (**Figure 9c**) of Mn-doped perovskites. Mn doping enhanced the luminescence of LEDs by 1.3 times of the pure PLEDs and EQE and CE also improved from 0.81% and 3.71 cd/A for the $CsPbBr_3$ PLED to 1.49% and 6.40 cd/A for the 3.8 mol% Mn-doped $CsPbBr_3$ PLED device. Similarly, Liu group also fabricated lead substituted active PLEDs with doping of 33 mol% Sn^{4+} ion consisting of the same device structure as Zou reported [40]. They reported that the presence of Sn^{4+} helped in the easy injection of charge carriers that is responsible for small turn off voltage as well as large current density. $CsPb_{0.67}Sn_{0.33}Br_3$ PLEDs showed 4.13% EQE, 12500 cd m^{-2} luminescence, 11.63 cd A^{-1} of CE, 6.76 lm W^{-1} and 3.6 V turn-on voltage. Authors have also claimed that such kind of high performance is the best results found in Sn-based PLEDs. So this lead substituted Sn-based PQDs LED can be suitable for different active LEDs as well for backlight-based display devices. On the other hand, $CsPbCl_3$: Mn PQDs has also been fabricated in PLEDs with 2.2 lm W^{-1} of luminous intensity (**Figure 9d–f**) and good stability after continuous applying 3.5 V of voltage for 200 h [45]. 27 mol% Mn: $CsPbCl_3$ was mixed with curable resin and coated on 365 nm commercial UV-LED chip for the PL PLED device. Due to d-d transition, Mn doped $CsPbCl_3$ PLED gives bright orange emission with 54% PLQY in ambient conditions. In most of the optoelectronic devices, photo and thermal

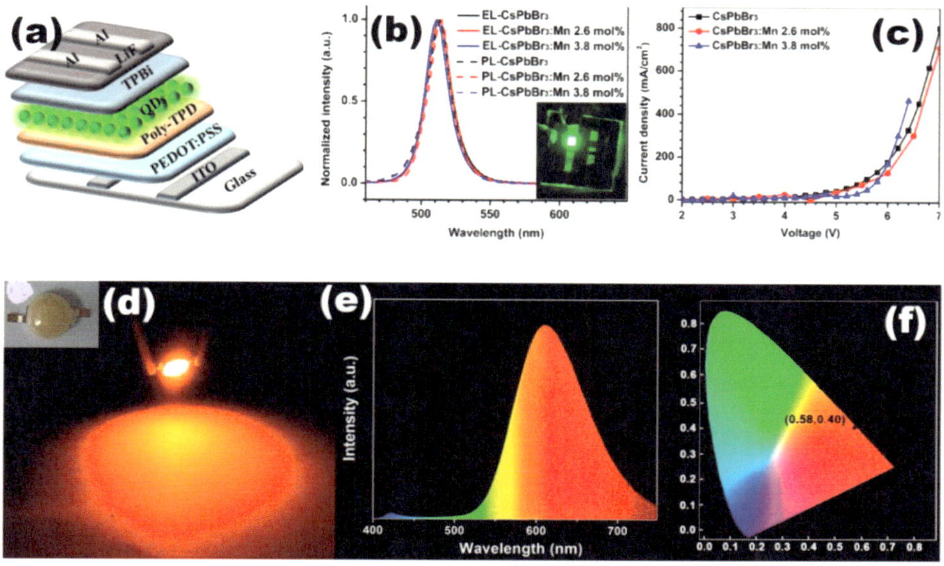

Figure 9.
(a) Schematic illustration of a typical multilayer-structured PLED device by using pure $CsPbBr_3$ and $CsPbBr_3$: Mn (2.6 mol%) and $CsPbBr_3$: Mn (3.8 mol%) QDs are used as green light emitters. (b) Comparison of normalized EL spectra at an applied voltage of 6 V and their corresponding PL emission spectra for $CsPbBr_3$ and $CsPbBr_3$: Mn (2.6 mol%) and $CsPbBr_3$: Mn (3.8 mol%) QDs when dispersed in cyclohexane solution. The inset shows a photograph of the EL of a representative PLED device. Current density (c) and luminance characteristics for three types of PLEDs based on the pure $CsPbBr_3$ and $CsPbBr_3$: Mn (2.6 mol%), and $CsPbBr_3$: Mn (3.8 mol%) QDs (d) fluorescent image (e), PL emission spectrum (f) CIE chromaticity coordinate of the LED from $CsPb_xMn_{1-x}Cl_3$ QDs. Inset: Optical image of the LED.

stability are the crucial issues. So for improving such kind of stability SiO_2 coating and KCl/polystyrene or other inorganic materials has been used for encapsulation of perovskite LEDs [42].

6. Summary

Lead halide-based perovskite material opens a new opportunity in optoelectric since 2012, because of their excellent optoelectronic properties and high power conversion efficiency within a brief time. The big challenge for the perovskite QDs is the presence of toxicity due to lead. Already many groups have worked on leadless or lead-free perovskite materials. 100% replacement of Pb will be complicated but with replacement up to 60–70%, Pb with non-toxic cation may help to the commercialization of less lead-containing perovskite materials. Shortly, less lead or lead-free PQDs have the potential for next-generation optoelectronic devices like backlights, QD TV, flexible and wearable devices, etc. However, to achieve high PLQY, good performances, and stability of lead-free or less lead containing PQDs also be challenging for the research community. Many homovalent lead substitutions like Sn, Mn, Cd, and Zn have been successfully done in place of Pb in different halide containing perovskite nanomaterials. Such kind of replacement of lead not only reduces the toxicity of perovskite while it also improves the performances and stability of the perovskite light-harvesting material. In the halide-based perovskites, dopant engineering is a compelling strategy to tuning the optical, structural, and electrical properties and day by day new depends arrive that make the PQDs more suitable for lead-free perovskite devices. Mn and Cd incorporation in Cl-based perovskite NCs was increased the PLQYs but for Br- and I-based PNCs it is still challenging to get high PLQYs than pure PQDs. So, we need to find a more suitable strategy for less lead-containing high PLQYs perovskite material.

Author details

Rajan Kumar Singh[1]*, Neha Jain[1], Sudipta Som[2], Somrita Dutta[3], Jai Singh[1] and Ranveer Kumar[1]

1 Dr. Harisingh Gour Central University, Sagar, MP, India

2 National Taiwan University, Taipei, Taiwan (Republic of China)

3 National Chiao Tung University, Hsinchu, Taiwan

*Address all correspondence to: rajanphysicssgo@gmail.com

References

[1] Noh JH, Im SH, Heo JH, Mandal TN, Seok SI. Chemical management for colorful, efficient, and stable inorganic-organic hybrid nanostructured solar cells. Nano Letters. 2013;**13**:1764-1769

[2] Kojima A, Teshima K, Shirai Y, Miyasaka T. Organometal halide perovskites as visible-light sensitizers for photovoltaic cells. Journal of the American Chemical Society. 2009;**131**:6050-6051

[3] Guerrero A et al. How the charge-neutrality level of interface states controls energy level alignment in cathode contacts of organic bulk-heterojunction solar cells. ACS Nano. 2012;**6**:3453-3460

[4] Yang X et al. Author correction: Efficient green light-emitting diodes based on quasi-two-dimensional composition and phase engineered perovskite with surface passivation. Nature Communications. 2018;**9**:1169

[5] Schmidt LC et al. Nontemplate synthesis of CH3NH3PbBr3 perovskite nanoparticles. Journal of the American Chemical Society. 2014;**136**:850-853

[6] Kovalenko MV, Protesescu L, Bodnarchuk MI. Properties and potential optoelectronic applications of lead halide perovskite nanocrystals. Science. 2017;**358**:745-750

[7] Gonzalez-Carrero S, Galian RE, Pérez-Prieto J. Maximizing the emissive properties of CH3NH3PbBr3 perovskite nanoparticles. Journal of Materials Chemistry A. 2015;**3**:9187-9193

[8] Moyen E et al. Ligand removal and photo-activation of CsPbBr3 quantum dots for enhanced optoelectronic devices. Nanoscale. 2018;**10**:8591-8599

[9] Hong K, Van Le Q, Kim SY, Jang HW. Low-dimensional halide perovskites: Review and issues. Journal of Materials Chemistry C. 2018;**6**:2189-2209

[10] Wei Y, Cheng Z, Lin J. An overview on enhancing the stability of lead halide perovskite quantum dots and their applications in phosphor-converted LEDs. Chemical Society Reviews. 2019;**48**:310-350

[11] Bella F et al. Improving efficiency and stability of perovskite solar cells with photocurable fluoropolymers. Science. 2016;**354**:203-206

[12] Wang R et al. A review of perovskites solar cell stability. Advanced Functional Materials. 2019;**1808843**:1-25

[13] Song T-B, Yokoyama T, Aramaki S, Kanatzidis MG. Performance enhancement of lead-free tin-based perovskite solar cells with reducing atmosphere-assisted dispersible additive. ACS Energy Letters. 2017;**2**:897-903

[14] Lyu M, Yun J-H, Chen P, Hao M, Wang L. Addressing toxicity of lead: Progress and applications of low-toxic metal halide perovskites and their derivatives. Advanced Energy Materials. 2017;**7**:1602512

[15] Babayigit A et al. Assessing the toxicity of Pb- and Sn-based perovskite solar cells in model organism danio rerio. Scientific Reports. 2016;**6**:18721

[16] Mitzi DB. $N_4H_9Cu_7S_4$: A hydrazinium-based salt with a layered Cu7S4-framework. Inorganic Chemistry. 2007;**46**(3):926-931

[17] Singh RK et al. Exploring the impact of the Pb^{2+} substitution by Cd^{2+} on the structural and morphological properties of $CH_3NH_3PbI_3$ perovskite. Applied Nanoscience. 2019:1-10. https://doi.org/10.1007/s13204-019-01021-5

[18] Kim HS et al. Lead iodide perovskite sensitized all-solid-state submicron thin film mesoscopic solar cell with efficiency exceeding 9%. Scientific Reports. 2012;**2**:591

[19] Singh RK et al. Novel synthesis process of methyl ammonium bromide and effect of particle size on structural, optical and thermodynamic behavior of $CH_3NH_3PbBr_3$ organometallic perovskite light harvester. Journal of Alloys and Compounds. 2018;**743**:728-736

[20] Chen Q et al. Under the spotlight: The organic–inorganic hybrid halide perovskite for optoelectronic applications. Nano Today. 2015;**10**:355-396

[21] Peña MA, Fierro JLG. Chemical structures and performance of perovskite oxides. Chemical Reviews. 2001;**101**:1981-2017

[22] Zhang F et al. Colloidal synthesis of air-stable CH3NH3PbI3 quantum dots by gaining chemical insight into the solvent effects. Chemistry of Materials. 2017;**29**:3793-3799

[23] Zhang S et al. Efficient red perovskite light-emitting diodes based on solution-processed multiple quantum wells. Advanced Materials. 2017;**29**:1606600

[24] Singh RK et al. Solution processed hybrid organic-inorganic CH3NH3PbI3 perovskite material and optical properties. Materials Today: Proceedings. 2017;**4**:12661-12665

[25] Price MB et al. Hot-carrier cooling and photoinduced refractive index changes in organic–inorganic lead halide perovskites. Nature Communications. 2015;**6**:8420

[26] Sheng R et al. Photoluminescence characterisations of a dynamic aging process of organic–inorganic CH3NH3PbBr3 perovskite. Nanoscale. 2016;**8**:1926-1931

[27] Tan Z-K et al. Bright light-emitting diodes based on organometal halide perovskite. Nature Nanotechnology. 2014;**9**:687

[28] Shamsi J, Urban AS, Imran M, De Trizio L, Manna L. Metal halide perovskite nanocrystals: Synthesis, post-synthesis modifications, and their optical properties. Chemical Reviews. 2019;**119**:3296-3348

[29] Hoefler SF, Trimmel G, Rath T. Progress on lead-free metal halide perovskites for photovoltaic applications: A review. Monatshefte für Chemie—Chemical Monthly. 2017;**148**:795-826

[30] Konstantakou M, Stergiopoulos T. A critical review on tin halide perovskite solar cells. Journal of Materials Chemistry A. 2017;**5**:11518-11549

[31] Shum K et al. Synthesis and characterization of CsSnI3 thin films. Applied Physics Letters. 2010;**96**:221903

[32] Chung I et al. CsSnI3: Semiconductor or metal? High electrical conductivity and strong near-infrared photoluminescence from a single material. High hole mobility and phase-transitions. Journal of the American Chemical Society. 2012;**134**:8579-8587

[33] Hou S, Guo Y, Tang Y, Quan Q. Synthesis and stabilization of colloidal perovskite nanocrystals by multidentate polymer micelles. ACS Applied Materials & Interfaces. 2017;**9**:18417-18422

[34] Fang G et al. Reverse synthesis of $CsPb_xMn_{1-x}(Cl/Br)3$ perovskite quantum dots from CsMnCl3 precursors through cation exchange. Journal of Materials Chemistry C. 2018;**6**:5908-5915

[35] Marshall KP, Walker M, Walton RI, Hatton RA. Enhanced stability and efficiency in hole-transport-layer-free CsSnI3 perovskite photovoltaics. Nature Energy. 2016;**1**:16178

[36] Liu F et al. Colloidal synthesis of air-stable alloyed $CsSn_{1-x}Pb_xI_3$ perovskite nanocrystals for use in solar cells. Journal of the American Chemical Society. 2017;**139**:16708-16719

[37] Wang A et al. Controlled synthesis of lead-free and stable perovskite derivative Cs_2SnI_6 nanocrystals via a facile hot-injection process. Chemistry of Materials. 2016;**28**:8132-8140

[38] Eames C et al. Ionic transport in hybrid lead iodide perovskite solar cells. Nature Communications. 2015;**6**:7497

[39] Wang H-C et al. High-performance $CsPb_{1-x}Sn_xBr_3$ perovskite quantum dots for light-emitting diodes. Angewandte Chemie, International Edition. 2017;**56**:13650-13654

[40] Akkerman QA et al. Tuning the optical properties of cesium lead halide perovskite nanocrystals by anion exchange reactions. Journal of the American Chemical Society. 2015;**137**:10276-10281

[41] Buonsanti R, Milliron DJ. Chemistry of doped colloidal nanocrystals. Chemistry of Materials. 2013;**25**:1305-1317

[42] Zhao J. Enhanced luminescence and energy transfer in Mn2+ dopedB-site doped $CsPbCl_3$-xBrx perovskite nanocrystals. Nanoscale. 2018;**10**:19435-19442

[43] Guria AK, Dutta SK, Das Adhikari S, Pradhan N. Doping Mn^{2+} in lead halide perovskite nanocrystals: Successes and challenges. ACS Energy Letters. 2017;**2**:1014-1021

[44] Liu F et al. Highly luminescent phase-stable CsPbI3 perovskite quantum dots achieving near 100% absolute photoluminescence quantum yield. ACS Nano. 2017;**11**:10373-10383

[45] Liu W et al. Mn^{2+}-doped lead halide perovskite nanocrystals with dual-color emission controlled by halide content. Journal of the American Chemical Society. 2016;**138**:14954-14961

[46] Yuan X et al. Photoluminescence temperature dependence, dynamics, and quantum efficiencies in Mn^{2+}-doped $CsPbCl_3$ perovskite nanocrystals with varied dopant concentration. Chemistry of Materials. 2017;**29**:8003-8011

[47] Zhu J et al. Room-temperature synthesis of Mn-doped cesium lead halide quantum dots with high Mn substitution ratio. Journal of Physical Chemistry Letters. 2017;**8**:4167-4171

[48] Zou S et al. Stabilizing cesium lead halide perovskite lattice through Mn(II) substitution for air-stable light-emitting diodes. Journal of the American Chemical Society. 2017;**139**:11443-11450

[49] Akkerman QA, Meggiolaro D, Dang Z, De Angelis F, Manna L. Fluorescent alloy $CsPb_xMn_{1-x}$I3 perovskite nanocrystals with high structural and optical stability. ACS Energy Letters. 2017;**2**:2183-2186

[50] van der Stam W et al. Highly emissive divalent-ion-doped colloidal $CsPb_{1-x}M_xBr_3$ perovskite nanocrystals through cation exchange. Journal of the American Chemical Society. 2017;**139**:4087-4097

[51] Mondal N, De A, Samanta A. Achieving near-unity photoluminescence efficiency for blue-violet-emitting perovskite nanocrystals. ACS Energy Letters. 2019;**4**:32-39

Chapter 8

Perovskite Materials for Resistive Random Access Memories

Jiaqi Zhang and Wubo Li

Abstract

Resistive random access memory (RRAM) utilizes the resistive switching behavior to store information. Compared to charge-based memory devices, the merits of RRAM devices include multi-bit capability, smaller cell size, and energy per bit (~fJ/bit). In this chapter, we review different perovskite material-based resistive random access memories (RRAMs). We first introduce the history of RRAM development and operational mechanism of conduction, followed by a review of two types of materials with perovskite crystal structure. One is conventional perovskite oxides (PCMO, a-LCMO, etc.), and the other is perovskite halides (organic-inorganic hybrid perovskites and inorganic perovskites) that have recently emerged as novel materials in optoelectronic fields. Our goal is to give a comprehensive review of perovskite-based RRAM materials that can be used for neuromorphic computing and to help further ongoing development in the field.

Keywords: memory devices, RRAM, perovskite oxides, perovskite halides, memristors

1. Introduction

Resistive random access memory (RRAM) devices use the resistance switching (RS) behavior to save information, which has attracted much attention due to its outstanding performance compared with traditional semiconductor electronic devices. In comparison to charge-based memory cells, RRAM device possesses various advantages, such as multi-bit capability, simpler device structure (electrode/active layer/electrode), as well as lower energy consumption (~fJ/bit) [1]. Another advantage of RRAM is its good compatibility to the conventional CMOS, which allows it to be integrated into current integrated circuit (IC) technology [2].

RRAM has various potential applications. First, RRAM is considered as the most promising candidate as the next-generation memory device because it acts excellently as both main memory and working memory. As main memory, RRAM is nonvolatile with high memory capacity. As working memory, the operation voltage and power of RRAM are very low, and the write/erase rate is very high. Apart from the memory application, RRAM is also utilized in low-energy-consumption computing as nonvolatile logic circuit [3, 4] and in neuromorphic computing as a synaptic cell, with the latter being a research hotspot in recent years [1].

A typical RRAM device is a metal/insulator/metal (MIM) stack, in which an insulating active layer is sandwiched between the top and the bottom metal electrodes, as shown in **Figure 1**. The device resistance can be tuned by applying an

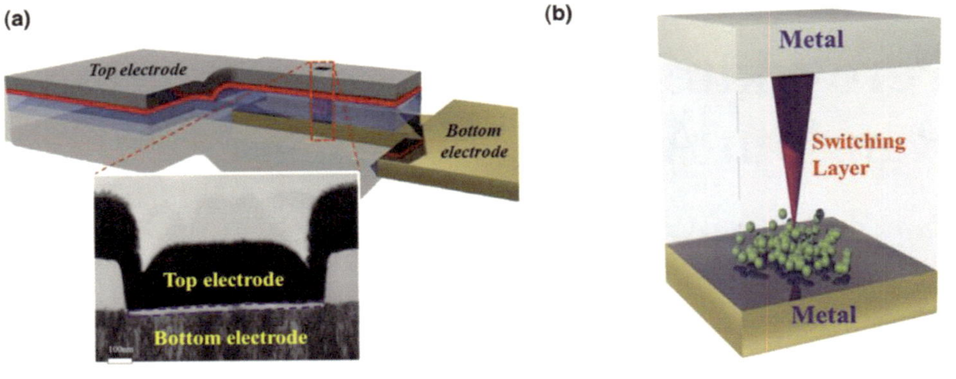

Figure 1.
(a) Schematic diagram of an RRAM device; (b) cross-sectional view of a RRAM device with conductive filament mechanism. Reproduced with permission [2]. Copyright 2016, Elsevier.

electric bias across two electrodes, forming high resistance state (HRS) and low resistance state (LRS). So the nonvolatile memory phenomenon in RRAM device is realized by electrically modulating the RS between HRS and LRS.

The transformation from HRS to LRS is normally named the Set process, and the transformation from LRS to HRS is named the Reset process. There are three kinds of RS switching behaviors. One is unipolar switching, in which the Set and Reset processes can happen at the same polarity of the external bias. Second is bipolar switching, in which RS switching occurs at different polarities of the applied bias. The third is nonpolar resistive memory, in which both RS switching from HRS to LRS and RS switching from LRS to HRS can be achieved without altering the voltage polarity (unipolar), while they can also be achieved by altering the polarity (bipolar).

The schematic I-V characteristics of unipolar switching and bipolar switching are illustrated in **Figure 2**, respectively. In addition to the Set and Reset processes, a Forming process typically exists for many initially prepared RRAM devices, in which an applied forming voltage (V_{form}) drives the formation of CFs with the compliance current limitation. The forming process is normally accomplished before the RRAM device enables to work, and the V_{form} is usually larger than the setting voltage (V_{set}).

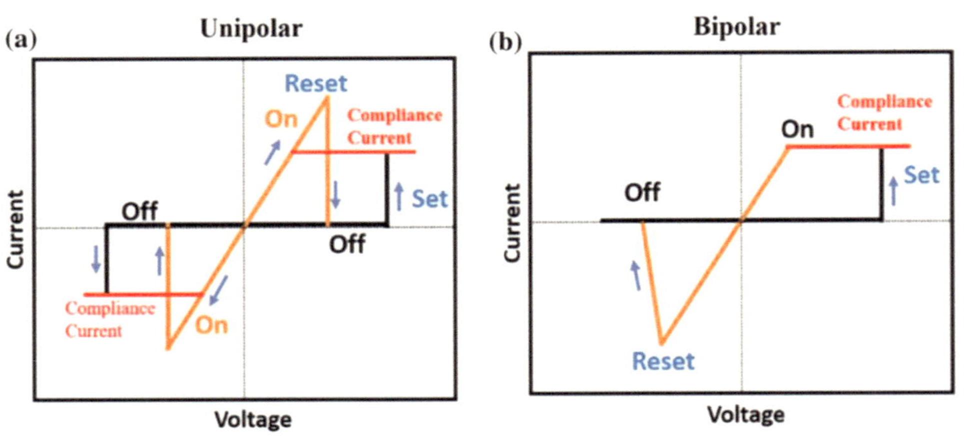

Figure 2.
Typical current voltage (I-V) characteristics of memory devices with (a) unipolar switching and (b) bipolar switching. Reproduced with permission [1]. Copyright 2018, Springer.

As aforementioned, a basic RRAM device usually consists of two electrodes and an active layer. The top and bottom electrodes can use various materials, including elementary substantial metals (Ag, Cu, Al, Au, Pt, W, etc.) [5], metallic alloys (Pt-Al, Cu-Ti, etc.) [6], and oxides (ITO, $SrRuO_3$, Nb:$SrTiO_3$, etc.) [7–9]. Based on the functions in RS conversion, the electrode materials can be divided into two types. One is active electrodes (Cu, Ag, etc.), which contribute to the RS conversion by the migration and/or redox reaction of the electrode ions around the electrode/active layer junction. The other is inert electrodes (Pt, Au, etc.), which do not directly participate in the RS conversion.

As for the active layer, various materials have been utilized in memory devices, such as amorphous metal oxides, polymers, hybrid composites, perovskite oxides, and perovskite halides [10, 11]. The material choice of RRAM active layer has significant influence on the device performance. In this chapter, we will mainly introduce oxides and halides with perovskite structures. Perovskite oxides are a conventional active material family for memories. In addition to high-endurance, chemically stable, and high-speed operation, the strong electron correlation induces many unique properties for perovskite oxides, which makes it remain as one of the most promising materials for RRAM yet. Compared with conventional RRAMs, perovskite halides can make flexible devices with low-cost fabrication, compositional flexibility, and excellent optoelectronic properties, enabling the perovskite halides as a promising next-generation memory material family. Next, we will introduce the memory devices with perovskite oxides and halides separately.

2. Perovskite oxide memory devices

In 2002, perovskite oxide $Pr_{0.7}Ca_{0.3}MnO_3$ was firstly utilized in a 64-bit RRAM array by a 500-nm complementary metal oxide semiconductor (CMOS) process [1]. Nearly two decades passed, transition metal perovskite oxides are still one of the best materials for RRAM active layers. Perovskite oxides possess stable crystal structure with high defect tolerance and structure flexibility, which enables the accommodation of nonstoichiometric ions. The nonstoichiometric ions contribute to the local ionic migration and thermochemical reaction therefore allowing for the RS conversion. In addition, the strongly correlated electrons in perovskite oxides provide various electronic phases and bring out multifunctionality, e.g., colossal magnetoresistance/electroresistance, ferroelectricity, multiferroics, superconductivity, etc. [12]. Also, the competing behavior among different electronic phases brings out the metal-insulator transition (MIT) phenomenon, which allows a significant change of resistance with a tiny electric stimulus [12, 13]. In the following, based on different resistance switching mechanisms, we will introduce RRAM with transition metal perovskite oxides via filament mode and uniform mode, respectively.

2.1 Conductive filament mechanism

Conductive filament (CF) is the most common mechanism to explain the resistance switching of memristor devices, in which the formation and the breakage of filaments are in Set and Reset process, respectively. Many works have proven the conductive filament in RRAM devices. For instance, a co-doped $BaTiO_3$-based device is forming as LRS, and the top electrode is divided into two portions (TE-I and TE-II). Then people measured the resistance between the bottom electrode and TE-I or TE-II. Significantly different resistances are found for two parts, which indicates inhomogeneous conductivity inside the whole memory device [14]. Besides, in an $YMnO_3$-based memory device, people found that the switching I-V curves of devices with various electrode areas are not remarkably different and that the lateral distribution of filaments is not uniform [15]. Real-time observation of conductive filament was also conducted by TEM in oxide-based memory devices, which shows direct evidence of conductive filament mechanism [16, 17].

Next, we will give a brief introduction on the microscopic mechanisms of filaments, mainly including two types: ion migration and metal-insulator transition.

2.1.1 Ion migration

For the conductive filaments caused by ion migration, the presence of CFs is very random owing to the random distribution of ions and defects in the active layer. Therefore, the formation and rupture of the CFs are strongly dependent on the initial distribution of ions and defects. For bipolar RRAM devices, the microscopic mechanism of ion migration can be classified into three types.

First, the filament is formed by the local redox reaction of metallic cations from the active metal electrode. For example, in an Ag/a-LSMO (amorphous Sr-doped $LaMnO_3$)/Pt device, Ag is demonstrated as the main component of the CFs [18]. When a positive bias is applied to the Ag electrode, some Ag atoms around the Ag/a-LSMO interface will be oxidized into Ag cations, and these Ag cations will move toward the opposite cathode under the external electrical field and are eventually reduced back to Ag atoms around the cathode/a-LSMO interface, therefore forming the CFs between two electrodes. Now if we apply a negative bias to the Ag electrode, the Ag atoms around the cathode will be re-oxidized into Ag^+ and move back toward the Ag electrode, thus leading to the breakage of the filaments. A typical RRAM device of this mode is an insulating active layer sandwiched between an active metallic electrode and an inert electrode [17].

A second mechanism of ion migration is that the positive-charged defects drift under the external voltage. The charged defects, such as oxygen vacancies and excess cations, can tune the Fermi level and correspondingly change the electrical conductivity in the local area. For instance, in a TiO_2-based memory device, the filaments are formed by the oxygen-deficient Ti_4O_7 phase under positive voltage [19]. When applying a negative voltage, the reverse redox occurs with the backward electric field and the parasitic Joule heating and consequently leads to the rupture of CFs [20].

Redox reactions both exist in the aforementioned two microscopic mechanisms. However, bipolar RS phenomenon also can be caused by the ion migration without redox. Pt/NSTO (Nb-doped $SrTiO_3$) device is taken as an example. Under electrical field, the movement of oxygen vacancies can change the Schottky barrier height and the depletion width of the Pt/NSTO junction at some local areas of the interface, resulting in the change of the electrical conductivity alongside the Pt/NSTO interface [21, 22].

For unipolar RS behavior with ion migration, the rupture of the CFs is different from that in the bipolar counterparts. In bipolar devices, the filament rupture is caused by the retraction of the initially moved ions or by the change of interfacial junction barriers. However, in unipolar memory devices, the filament rupture is driven by the Joule heat-assisted thermochemical reaction. For example, in a Au/$YMn_{1-\delta}O_3$/Pt unipolar memory device, after the filament is formed under forward bias, a reverse bias with a similar value cannot supply sufficient energy to retract the initially migrated ions and activate the local reverse redox [15]. Instead, the electrical current can provide enough Joule heat in local areas of the filament; sometimes the local temperature can be increased by several hundreds of Kelvin [15, 23], thus assuring the local reverse redox and the corresponding rupture of CFs. As for the HRS to LRS transformation, accompanied by the ion migration, the electron hopping barriers and the related trapping states which exist in HRS are removed by the further increase of the applying voltage [15, 17, 24]. The same mechanism has also been proven in the Au/co-doped $BaTiO_3$/Pt unipolar memory device [14].

2.1.2 Metal-insulator transition

Metal-insulator transition (MIT) effect has been found in many perovskite oxides, in which the electronic charges are injected into the insulating material to induct the current with an external bias [25–27]. $Pr_{1-x}Ca_xMnO_3$ (PCMO), now one of the most developed memory materials for neuromorphic computing, is taken as an example, in which resistive switching behavior was first discovered in late 1990s [25]. The electron injection distorts the superlattice structure and the mixed valence band in the strongly electronic correlated PCBM system, which acts as an ion doping process. The rebuilding of the electronic phase separation state can also contribute to the MIT, induced by the external electrical stimulus and parasitical Joule heating, which exhibits CFs-based unipolar RS phenomenon [28]. In addition, filament-type RS behavior may also derive from the Mott transition, which has been demonstrated in many transition metal perovskite oxides [26, 29, 30].

2.2 Uniform resistance switching

In CF-based memory devices, the conductive filaments are formed under electric stimuli in local areas. The I-V characteristics are not proportional to the electrode area due to the random distribution of the filaments. Apart from the CF-based RRAMs, uniform resistance switching mechanism has already been demonstrated, in which the device resistance variation is spatially uniform. Thus the variations of HRS and LRS are both proportional to the electrode area. Uniform RS behavior mainly includes two types, one is the carrier trapping/detrapping, and the other is the ferroelectric polarization.

Carrier trapping/detrapping and the migration of charged defects can tune the Schottky barrier at the metal-insulator interface thus modulating the device resistance. This modulation could occur in local regions near the interface (i.e., filament mode) or occur laterally uniform near the interface (i.e., uniform mode), which is strongly dependent on the interfacial electrical and morphological uniformity. A smooth interface with uniform distribution of charges and defects may bring out uniform RS. Otherwise, filaments may tend to form with nonuniform interfaces. Researchers found that the uniform migration of charged defects (e.g., oxygen vacancies) is too slow thus leading to very slow RS [31]. Nevertheless, the charge trapping/detrapping at the junction can be very fast, enabling uniform RS with fast response, which has been confirmed in the Au/Nb-doped $SrTiO_3$ heterojunction [14].

For the uniform RS by ferroelectric polarization, ferroelectric tunnel junctions (FTJs) are utilized for the RRAM devices, including a ferroelectric tunnel barrier sandwiched by two electrodes. Many perovskite oxide materials have been utilized in the FTJ-based memory devices, such as Pt/$BaTiO_3$/$SrRuO_3$ [8], Pt/$BiFeO_3$/$SrRuO_3$ [32], Co/$BaTiO_3$/$La_{0.7}Sr_{0.3}MnO_3$ [33], etc. The polarization at the ferroelectric/metal junction has a significant influence on the junction barrier profile and modulates the electron tunneling. Thus when the polarization is varied with the external electric field, the resistance state is obviously changed.

Although uniform RS behavior exhibits many advantages, currently the practical application of uniform-type memory devices is still restricted by some intrinsic demerits. The key issue is that the uniform-type device performance is closely dependent on the quality of the films and the junction [6]. For the carrier trapping-/detrapping-based devices, the LRS and the HRS often show considerable relaxation, which deteriorates the device performance [9]. An effective interfacial modification is commonly required to solve this problem. For the FTJ-based devices, the tunneling current tuned by the polarization is normally remarkably small, which hinders its actual application. In addition, the ferroelectric layer is usually ultrathin, and how to maintain the ultrathin film uniformity in a large scale is another technical issue.

3. Perovskite halide memory devices

In recent years, halide perovskites (HPs) have become a star material due to its excellent optical and charge transport properties. The rapid advance in power-conversion efficiency (PCE) of perovskite solar cells has exceeded by 20% [34], and the simple and solution-based preparation enables low-cost production. HPs have excellent electron migration ability and good optical absorption. With the development of HPs, the hysteresis in the current voltage curve was observed and described [35]. It is found that the hysteresis has a strong dependence on the voltage scanning rate and transient response. Ion migration is thought to be a possible origin of the slow response [36]. This discovery paves the way for HPs' applications in other electronic devices, for example, resistive switching memory (memristors) [37–39], field-effect transistors [40–42], and artificial synapse devices [43, 44]. Owing to its unique features and manufacturing advantages, rapid progress has been made, and HPs are considered as a promising candidate for the next generation of electronic devices [45, 46].

3.1 Tunable bandgap

The perovskite material is an ABX_3 compound with a 3D framework (**Figure 3a**), where A is a monovalent cation. A-site can be an inorganic or organic cation, for example, methylammonium (MA^+, $CH_3NH_3^+$), formamidinium (FA^+, $HC(NH_2)^{2+}$), or Cs^+; B is a divalent cation, and X is an anion. B is typically Pb (also Sn) and X is a halide such as Cl, Br, or I. Based on the composition flexibility of

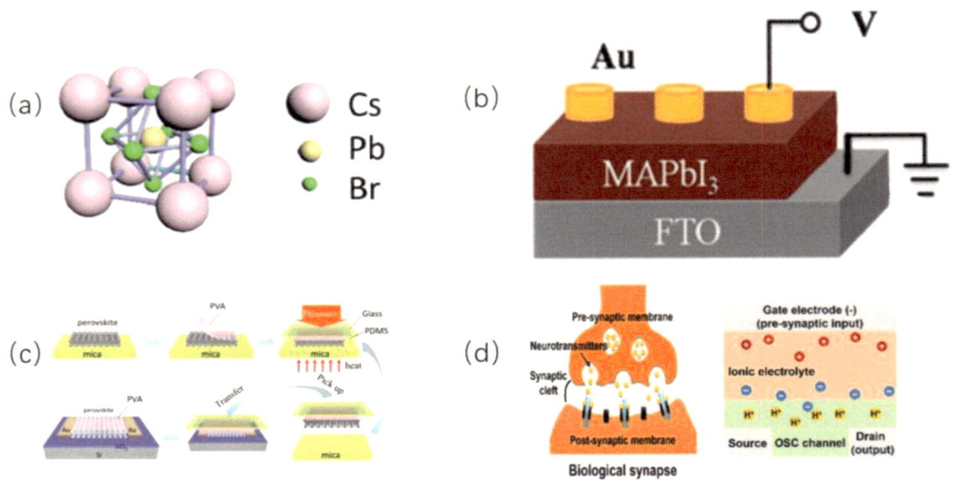

Figure 3.
(a) Crystal structure of HPs [47]. (b) A schematic diagram of device structure halide perovskite memory [48]. (c) FETs based on $CsPbBr_3$ and fabrication method [47]. (d) Biological synapse compared to artificial synapses [49]. (a, c) Reproduced with permission [47]. Copyright 2015, American Chemical Society. (b) Reproduced with permission [48]. Copyright 2018, American Chemical Society. (d) Reproduced with permission [49]. Copyright 2019, American Chemical Society.

HPs, the bandgap can be tuned by replacing elements at each position. In addition, the bandgap can be tuned by controlling the crystalline structure of HPs and the grain size [50].

It is a feasible method to change the length of the bonds between A and B/C sites. In theory, the crystal lattice of the perovskite ABX_3 is expandable, and the gap of the forbidden band is narrow. For example, the material obtained by replacing the MA^+ in the $MAPbI_3$ with an ethylamine, a propylamine, a long-chain alkyl, or an arylamine cation is generally a two-dimensional layered structure. The length of alkylammonium cations at position A was reported in 1990 by Calabrese et al., and the synthesized HPs demonstrated a two-dimensional layered structure [51]. With the increase of the length of cation at site A, the maximum absorption peak is red-shifted from 390 to 450 nm [51]. The modulation of the perovskite bandgap by the substitution of A-site has been demonstrated by the density functional theory (DFT), showing an obvious change, i.e., FA (1.5 eV), MA (1.55 eV), and Cs (1.73 eV) [52, 53]. The angle of B-X-B bond in perovskite structure plays an important role in regulating the bandgap of perovskite materials. Therefore, the change of different metal ions (B) to regulate the structure and properties of perovskite materials is also of great concern. By substituting Pb by Sn at B site, $MASnI_3$ (1.3 eV) exhibits a smaller bandgap than $MAPbI_3$ (1.55 eV) [53]. For the X site, when I ions in $MASnI_3$ are doped with Br in different proportions, the bandgap of the materials can be modulated between 1.3 and 2.15 eV, and the corresponding absorptions are between 950 and 650 nm [54].

Another important way to adjust the bandgap is to control the quantum confinement in the nanoscale. Compared with 0D quantum dots or 1D nanowires, 2D geometry provides a natural way to accurately control the thickness of quantum wells for perovskite halides, resulting in a confinement effect. Huang et al. found that a two-dimensional $MAPbBr_3$ perovskite layer could be regulated by the concentration of oleic acid and the balance between surfactant and precursor in two phases [55]. Two-dimensional $MAPbBr_3$ nano-sheets with different layers show different representative absorption spectra and photoluminescence spectra [55].

3.2 Ion migration

Many perovskite photovoltaic cells have exhibited I-V hysteresis behavior, as shown in **Figure 4** [36, 56, 57]. Ion migration is thought to be an origin of the photocurrent hysteresis. Low formation energy of ionic defects combined with low activation energy for ion migration enables easy and fast ion migration in perovskite halide materials. Although raising potential stability is an issue in HP solar cells, the ion migration combined with the excellent optical and electrical properties of the material also provides an opportunity for new devices such as optically controlled memory and switched diodes.

HPs are good ionic conductors with fast ion migration ability. There are many factors affecting ion migration, such as component ions, defects, cation rotation, etc. We first briefly introduce the defect ions in perovskite crystals. HPs possess various intrinsic point defects, such as vacancies, interstitial defects, and antisite defects. Shao et al. found that ion migration at grain or grain boundary of $MAPbI_3$ perovskite membrane is different [58]. Ion migration in perovskite membranes can be regulated by the introduction of other foreign substances into grain boundaries, such as large fullerene derivatives ($PC_{60}BM$) or small chloride ions [59, 60]. The modulation of ion migration is desirable for the development of high-performance perovskite-based optically adjustable resistors and synaptic devices [60]. Perovskite has been proven as an excellent ion conductor. Because of the ion motion, the semiconductor material can be changed from p-doped to n-doped.

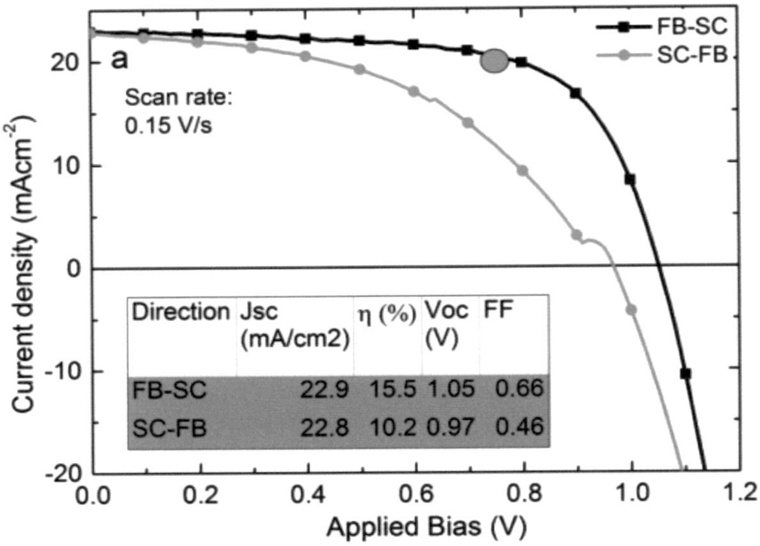

Figure 4.
A typical I-V hysteresis behavior of perovskite solar cells, with forward bias to short circuit sweep (FB-SC) and short circuit to forward bias sweep (SC-FB) [36]. Reproduced with permission [36]. Copyright 2014, American Chemical Society.

By applying external bias, the device structure can be changed from p-i-n structure to n-i-p structure, thus gradually changing the resistance of the device. This memory characteristic of perovskite materials can simulate the signal processing, learning, and memory functions of the nervous system [61]. Perovskite memristors can reduce the energy consumption required for the primary signal transmission of artificial synaptic devices to femto-Joule/(100 nm)^2 which is similar to the ultralow energy consumption required for primary signal transmission in biological synapses. Due to the excellent optical and electrical properties of perovskite materials, some biological functions read by optical signals have also been discovered [62].

3.3 Flexibility

Flexible devices have enormous potentials for applications in emerging areas such as wearable electronics, portable chargers, remote power supplies, automobiles, and aircrafts. The fabrication of the substrate is very important for the flexibility of the final device, and the flexible device based on the polymer substrate is usually needed, resulting in general processing, and manufacturing only in low-temperature environments cannot withstand high-temperature processes. But HP materials do not require high temperatures and can be processed at low temperatures, and HPs provide mechanical flexibility. These make HPs a great advantage in flexible device applications (**Figure 5**). For typical HPs, $MAPbBr_3$, they have weak interactions between organic elements. This combination is relatively weak, so the shear between perovskite surfaces is easy to occur, which explains why this perovskite can provide elasticity under mechanical deformation. The annealing temperature of HPs is generally only one hundred degrees. Therefore, high flexible polymer substrates can be used in HP-based flexible devices because of the low processing temperature. Many repeated bending tests of HP solar cells and storage devices have been reported. These studies show that the materials have a good mechanical flexibility.

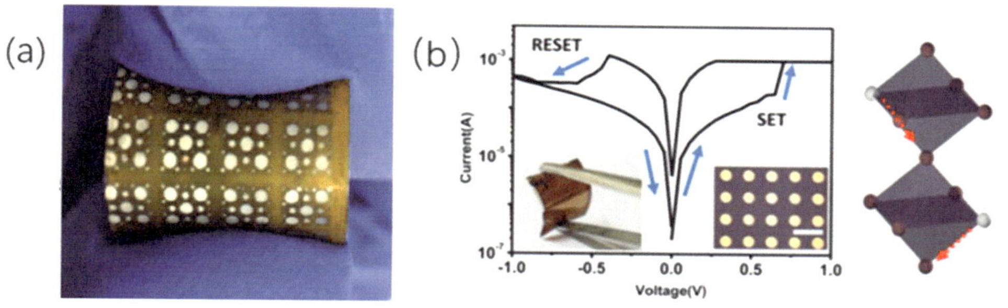

Figure 5.
(a) Photograph of a flexible RRAM device with the Al/$CsPbBr_3$/PEDOT:PSS/ITO structure [63]. (b) Memory device based on flexible substrate of the Au/perovskite/ITO structure and I-V characteristics [37]. (a) Reproduced with permission [63]. Copyright 2017, American Chemical Society. (b) Reproduced with permission [37]. Copyright 2016, American Chemical Society.

HP-based flexible resistive switch storage device has been fabricated on a plastic substrate. After more than 100 times of bending radius of 1.5 cm, the storage device still has electrical performance (**Figure 5b**) [37]. The first fiber-shaped perovskite memristor was developed in 2016 [64]. In particular, fiber morphology is expected to promote the application of perovskite materials in wearable memory and computing device. Therefore, thanks to the good mechanical and electrical reliability, HP-based devices are very promising for the next-generation flexible memory devices.

3.4 Thin film preparation methods

In the field of perovskite solar cells, it is necessary to improve the photoelectric conversion efficiency of perovskite solar cells with good thin film preparation technology [65, 66]. Generally good perovskite thin films have smooth surface and large grain size with relatively few defects. In order to achieve high-performance devices, controlling the uniformity, thickness, and grain size of the HP layer is of great importance in the fabrication process.

In the early preparation of the solar cell, the perovskite precursor solution is usually spin-coated on the hydrophilic TiO_2 layer, and due to the hydrophilicity, the perovskite is easily deposited on the TiO_2 [67, 68]. Resistive switch memory and logic device is a kind of device with an insulating layer sandwiched between two metal layers. Thus when we use a solution method to prepare HP layers for the memristor devices, the perovskite precursor solution needs to be deposited on the hydrophobic metal layer. However, it is difficult to use the solution method to deposit a HP film on a hydrophobic metal surface. Because of the hydrophobicity of metal electrodes, for example, a simple spin coating method of $MAPbI_3$ precursor solution may produce island growth on the metal surface. One-step spin coating therefore is not suitable for the fabrication of memory and logic device structures without interfacial modification. In order to solve this problem, the surface of metal electrode is usually treated with ultraviolet ozone (UVO) or O_2 plasma to change the hydrophobicity of metal electrode. However, it is still not easy for HP thin films to get a good uniformity, both in one-step spin coating and two-step spin coating. This mainly originates from the difference between the general perovskite layer annealing temperature and the solvent boiling point temperature. For example, the annealing temperature of $MAPBI_3$ is generally 100–150°C, while for the solvent perovskite, such as γ-butyrolactone and N,N dimethylformamide, possessing high boiling points of 204 and 153°C, respectively [44]. Thus the nucleation during the substrate annealing could be very slow, which tends to achieve poor film morphology, such as cracks or even pores. Anti-solvent engineering can be applied in spin coating process to eliminate this issue. In anti-solvent engineering, toluene, chloroform, and other substances are often used as anti-solvents. Because they are insoluble to perovskite, when anti-solvent is added, the anti-solvent begins to diffuse and permeate into HPs solution. It is helpful for rapid nucleation. Anti-solvent engineering has been successfully used in the fabrication of HP-based flexible resistive switch memory [37]. However, the use of anti-solvent engineering will also bring some problems. With the addition of anti-solvent, it gradually begins to diffuse and permeate in HP solution. However, it is not possible for the anti-solvent to diffuse and penetrate uniformly throughout the perovskite film, which may result in a large distribution of the perovskite crystal size throughout the film. In order to prepare more uniform membranes, it is usually necessary to add additives such as alkane dimercaptan to control the crystallization kinetics of perovskite [69].

3.5 HPs for resistive switching memories

With the advent of the information age and the rapid development of the Internet, the information that needs to be stored has been explosively increased, and the traditional storage equipment is more and more difficult to meet the demand. As a new-generation storage device, the memristor has great potential in the field of storage. In terms of storage performance, excellent memory devices need to have the advantages of fast working time, long service life, low power consumption, and low cost.

For memristor applications, many materials have been used, from organic materials and binary metal oxides to perovskite halide. Among them, metal oxide-based resistive switch devices have been extensively studied and applied in many fields. However, the technology has many demerits, such as high-power consumption and complicated fabrication process, which is not suitable for fabrication of flexible/wearable devices. As discussed above, perovskite halides are an ideal alternative to fabricate flexible devices [46].

For example, the change in the resistance switching for the $MAPbI_3$ memristor is a filament-type mechanism with the direct reaction of the charge carriers with the defects [39]. For RRAMs, fast charge transfer can reduce energy consumption. In HPs, the carrier transport capacity can be enhanced with appropriate concentrations of defects. For instance, doping $MAPbI_3$ with Br reduces the "SET" voltage, thereby reducing the power consumption of the device. This is because the activation energy of ion migration with Br vacancies is smaller than that with I. Thus, the HRS to LRS switching energy is reduced, and the switching response is accelerated.

At mentioned above, anti-solvent engineering has been utilized in the preparation of HP thin Films. The $MAPbI_3$ thin films treated with toluene as anti-solvent exhibit extremely low electric field about 3.25×10^3 V/cm and high switch-specific resistance switching behavior [70].

As a low-cost material, HPs have a great potential for the development of wearable and portable devices. Yan et al. developed the first fiber-shaped perovskite memristor [64]. In particular, fiber morphology is expected to promote the application of perovskite materials in wearable memory and computing device.

As the volume of information increases, devices that can store more data in the same size are the trend in the future. So, it is important to develop the memory device into a device with high storage density. Hwang et al. fabricated $MAPbI_3$ layer for nano-RRAM devices on 250 nm perforated silicon wafers by vapor deposition [71]. The device has the characteristics of bipolar resistance switch, low operating voltage, high switching speed (200 ns), high durability, and high data retention time (>10^5 s). In addition, the continuous vapor deposition technology is extended to $MAPbI_3$ memristor with a cross-point array structure. This method enables large area device fabrication for high-density memory devices [71].

All-inorganic perovskite halides, such as $CsPbBr_3$, have also demonstrated as working flexible nonvolatile memories, with a filament-type RS mechanism (**Figure 6a**) [63]. $CsPbBr_3$ quantum dots are also developed for the memory cells [73, 74]. Besides, due to lead in HPs being a component that pollutes the environment and is harmful to humans, it is also necessary to develop lead-free devices. Han et al. successfully fabricated RRAMs based on lead-free inorganic cesium iodide ($CsSnI_3$) perovskite material, as shown in **Figure 6b** [72]. Some typical HP-based RRAMs are compared, as shown in **Table 1**.

In addition to the traditional perovskite halides with ABX_3 structure, other new types of materials have also been utilized in memory devices. 2D perovskite is another promising candidate for RRAM. The conductivity of HPs is low, but it has a good carrier transport ability. At present, most of the HP RRAMs are based on 3D $MAPbX_3$ and some 2D Ruddlesden-Popper (RP) phase perovskite. 2D perovskite material has high Schottky barrier, 2D anisotropic structure, and electrothermal activation energy characteristics. Compared with 3D perovskite devices, the off current of 2D perovskite devices can be greatly reduced. Tian et al. reported the utilization of single-crystalline 2D $(PEA)_2PbBr_4$ and graphene for RRAM [76]. The two sides of 2D HPs are entrapped by graphene and Au, respectively. Due to the low conductivity of 2D HPs caused by multilayer organic ligands, there is no leakage current channel in perovskite grain boundaries of 2D HPs. The off current is limited to 1 pA. It is proven that the switching behavior has good reproducibility by switching devices at 10 pA program current circulate 100 times. Cheng et al. fabricated into Al/2D $(CH_3NH_3)_2PbI_2(SCN)_2$ perovskite film/indium-tin oxide [78]. The RRAM shows ternary switching. The three states have a conductivity ratio of $1{:}10^3{:}10^7$, with long retention over 10,000 s. A transparent 2D perovskite $(C_4H_9NH_3)_2PbBr_4$ has also been developed for compliance-free multilevel RS devices [79]. Ultrathin bismuth halide $Cs_3Bi_2I_9$ is also used as an electronic memory device with a typical bipolar RS behavior [80].

For most exploited devices, the data only transiently converts the optical signal into a circuit under illumination, which requires the use of additional converters to further store the output signal and record the occurrence of optical stimuli. HPs have a very strong optical absorption ability, low exciton binding energy, and long life carrier transmission time, so HPs can display a short signal under illumination, which can be used in light-stimulated devices. Chen et al. first introduced the concept of floating gate flash memory and successfully fabricated HP floating gate photomemory with a multilevel memory behavior [81]. Wang et al. first introduced a photonic RRAM based on $CsPbBr_3$ quantum dots. The $CsPbBr_3$ quantum dot layer

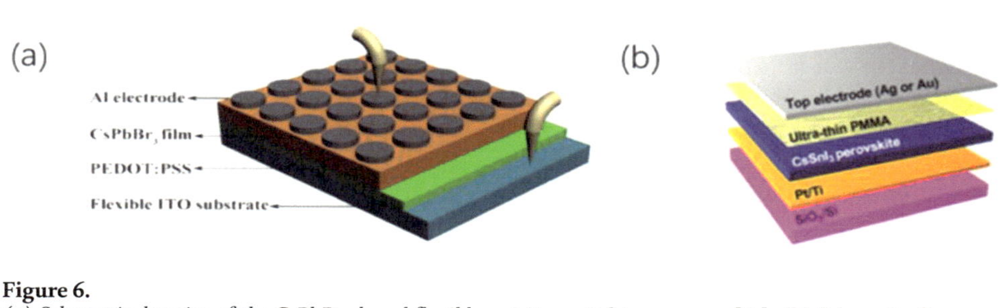

Figure 6.
(a) Schematic drawing of the $CsPbBr_3$-based flexible resistive switching memory [63]. (b) Schematic diagram of the Ag or Au/PMMA/$CsSnI_3$/Pt/SiO_2/Si vertical stack structure [72]. (a) Reproduced with permission [63]. Copyright 2017, American Chemical Society. (b) Reproduced with permission [72]. Copyright 2019, American Chemical Society.

Device structure	Set voltage [V]	On/off ratio	Retention [s]	Endurance [cycles]	Ref.
Au/$MAPbI_{3-x}Cl_x$/FTO	0.8	10	1×10^4	10^2	[38]
Ag/$MAPbI_3$/Pt	0.13	10^6	1×10^4	400	[70]
Ni/ZnO/$CsPbBr_3$/FTO	−0.95	10^5	1×10^4	—	[75]
Al/$CsPbBr_3$/PEDOT:PSS/ITO/PET	−0.6	10^2	—	50	[63]
Ag/PMMA/$CsPbI_3$/Pt	0.18	10^6	—	300	[62]
Graphene/PEA_2PbBr_4/Au	2.8	10	1×10^3	100	[76]
Ag/PMMA/$CsSnI_3$/Pt/SiO_2/Si	0.13	10^3	7×10^3	600	[72]
Au/$MAPbI_3$/Au	0.96	10^8	1×10^4	1000	[77]

Table 1.
Summary of hybrid perovskite RRAMs in this review.

is sandwiched by two PMMA layers. Silver is selected as the top electrode by thermal evaporation [82]. In the absence of light, the device displays a bipolar resistive switch memory. By inputting the light field and electric field signals, the current will be used as the output signal to realize the switching logic operation.

The traditional von Neumann architecture requires a large amount of data transmission directly in CPU and memory (memory wall). This leads to increased power consumption. In order to solve this problem, Tian et al. proposed a fully distributed architecture based on optical synapse. The optical synapse based on layered 2D $(PEA)_2PbI_4$ perovskite structure was prepared [83]. This 2D perovskite-type optical synapse is similar to the biologic optical synapse with light-induced excitation/inhibition. Based on the unique optical gate control effect, the ultrahigh light response rate can reach 730 a/w. Lead-free 2D perovskite was also utilized for the first time in the study of flexible optical synaptic devices [84]. A flexible optical synapse based on 2D perovskite $(PEA)_2SnI_4$ can mimic the short-term plasticity of biological synapses.

In addition to the single device operation, we should also pay attention to cross-array arrangement of RRAMs. A large number of RRAMs can be connected to each other in micro-space to form a cross-array structure. This architecture combines the memory advantage of the RRAMs and the massively parallel processing of the cross array. Cross arrays exhibit the characteristics of large-scale parallel processing, distributed information storage, self-organization, self-adaptation, etc. RRAM cross array provides a more convenient storage structure for binary images and a new storage scheme for gray-scale images. Hwang et al. prepared homogeneous perovskite thin films by sequential evaporation deposition and then prepared 16×16 cross-point array of RRAM [71]. The I-V characteristics of the memory cells show a variation among different points, while the setting voltages remain similar, and the on/off ratios are large for all devices. The memory characteristics prove the feasibility of HPs in the application of high-density cross-point memory. Kang et al. fabricated perovskite RRAM devices with high yield in 8×8 cross-bar arrays using solution-treated perovskite films [77]. Among the 64 memory cells, 55 cells are functional. These results are of great significance for the practical perovskite storage equipment with low cost and high density through a simple solution.

4. Conclusions

In this chapter, we have outlined an overview of the application of perovskite oxides and perovskite halides in memory devices. In the new era, artificial intelligence and IoTs are dramatically developing. Correspondingly, memory cells are getting more and more important, especially in low-power information storage and in neuromorphic computing. Although already developing for around two decades, perovskite oxides are still one of the most promising materials for RRAM owing to its high-endurance, chemically stable, and high-speed operation. However, more efforts are expected for perovskite oxide-based memories. Technologically, improving the endurance of the RS is still required for better actual application. Fundamentally, the basic operational mechanism of perovskite oxide RRAM device needs further investigation, especially considering the strong electron correlation system. For the perovskite halide, as a rising star, it has exhibited a great potential in the application of memristors. Flexible devices, low-cost fabrication, compositional flexibility, and excellent optoelectronic properties enable the perovskite halides to obtain potential application into wearable memory devices and artificial synapse. However, the film quality of HPs should be further improved, because the memory device performance is significantly dependent on the film uniformity. In addition, the intrinsic stability issue needs to be addressed by intended doping and interfacial passivation. Overall, further investigation is required to fulfill the expectation on these promising materials for the next-generation electronics.

Acknowledgements

JZ acknowledges the National Science Foundation of Jilin Province (grant: 20190201208JC), the Science and Technology Foundation of Department of Education, Jilin Province (grant: JJKH20190136KJ), and the Open Foundation of State Key Laboratory of Inorganic Synthesis and Preparative Chemistry, Jilin University (grant: 2019-24).

Conflict of interest

The authors declare that they have no conflict of interest.

Author details

Jiaqi Zhang* and Wubo Li
Key Laboratory of Automobile Materials of MOE, College of Materials Science and Engineering, Jilin University, Changchun, China

*Address all correspondence to: zhangjiaqi@jlu.edu.cn

References

[1] Hong XL, Loy DJJ, Dananjaya PA, Tan F, Ng CM, Lew WS. Oxide-based RRAM materials for neuromorphic computing. Journal of Materials Science. 2018;**53**:8720-8746. DOI: 10.1007/s10853-018-2134-6

[2] Chang TC, Chang KC, Tsai TM, Chu TJ, Sze SM. Resistance random access memory. Materials Today. 2016;**19**:254-264. DOI: 10.1016/j.mattod.2015.11.009

[3] Yang JJ, Strukov DB, Stewart DR. Memristive devices for computing. Nature Nanotechnology. 2012;**8**:13-24. DOI: 10.1038/nnano.2012.240

[4] Zhu D, Li Y, Shen W, Zhou Z, Liu L, Zhang X. Resistive random access memory & its applications in storage & nonvolatile logic. Journal of Semiconductors. 2017;**38**:1-13. DOI: 10.1088/1674-4926/38/7/071002

[5] Gao S, Zeng F, Song C, Pan F, Chen C. Recent progress in resistive random access memories: Materials, switching mechanisms, and performance. Materials Science & Engineering R: Reports. 2014;**83**:1-59. DOI: 10.1016/j.mser.2014.06.002

[6] Yan ZB, Liu JM. Resistance switching memory in perovskite oxides. Annals of Physics. 2015;**358**:206-224. DOI: 10.1016/j.aop.2015.03.028

[7] Zhang J, Wu X, Ma X, Yuan L, Huang K, Feng S. Amporphous La0.75Sr0.25MnO_3 thin film fabricated by pulsed laser deposition as a medium layer for semitransparent resistive random access memory. Chinese Journal of Inorganic Chemistry. 2018;**34**:784-790

[8] Wen Z, You L, Wang J, Li A, Wu D. Temperature-dependent tunneling electroresistance in Pt/$BaTiO_3$/$SrRuO_3$ ferroelectric tunnel junctions. Applied Physics Letters. 2013;**103**:132913. DOI: 10.1063/1.4823580

[9] Yan ZB, Liu JM. Coexistence of high performance resistance and capacitance memory based on multilayered metal-oxide structures. Scientific Reports. 2013;**3**:2482. DOI: 10.1038/srep02482

[10] Gao S, Yi X, Shang J, Liu G, Li R-W. Organic and hybrid resistive switching materials and devices. Chemical Society Reviews. 2019;**48**:1531-1565. DOI: 10.1039/C8CS00614H

[11] Liu D, Cheng H, Zhu X, Wang G, Wang N. Analog memristors based on thickening/thinning of Ag nanofilaments in amorphous manganite thin films. ACS Applied Materials & Interfaces. 2013;**5**:11258-11264. DOI: 10.1021/am403497y

[12] Jeong DS, Thomas R, Katiyar RS, Scott JF, Kohlstedt H, Petraru A, et al. Emerging memories: Resistive switching mechanisms and current status. Reports on Progress in Physics. 2012;**75**:076502. DOI: 10.1088/0034-4885/75/7/076502

[13] Waser R, Dittmann R, Staikov G, Szot K. Redox-based resistive switching memories-nanoionic mechanisms, prospects, and challenges. Advanced Materials. 2009;**21**:2632-2663. DOI: 10.1002/adma.200900375

[14] Yan Z, Guo Y, Zhang G, Liu JM. High-performance programmable memory devices based on Co-doped $BaTiO_3$. Advanced Materials. 2011;**23**:1351-1355. DOI: 10.1002/adma.201004306

[15] Yan ZB, Li SZ, Wang KF, Liu JM. Unipolar resistive switching effect in YMn1-δO_3 thin films. Applied Physics Letters. 2010;**96**:1-4. DOI: 10.1063/1.3280380

[16] Sun H, Liu Q, Li C, Long S, Lv H, Bi C, et al. Memory switching: Direct observation of conversion between threshold switching and memory

switching induced by conductive filament morphology. Advanced Functional Materials. 2014;**24**:5772. DOI: 10.1002/adfm.201470243

[17] Yang Y, Gao P, Gaba S, Chang T, Pan X, Lu W. Observation of conducting filament growth in nanoscale resistive memories. Nature Communications. 2012;**3**:732-738. DOI: 10.1038/ncomms1737

[18] Liu D, Wang N, Wang G, Shao Z, Zhu X, Zhang C, et al. Nonvolatile bipolar resistive switching in amorphous Sr-doped $LaMnO_3$ thin films deposited by radio frequency magnetron sputtering. Applied Physics Letters. 2013;**102**:134105. DOI: 10.1063/1.4800229

[19] Kwon D-H, Kim KM, Jang JH, Jeon JM, Lee MH, Kim GH, et al. Atomic structure of conducting nanofilaments in TiO_2 resistive switching memory. Nature Nanotechnology. 2010;**5**:148-153. DOI: 10.1038/nnano.2009.456

[20] Larentis S, Nardi F, Balatti S, Gilmer DC, Ielmini D. Resistive switching by voltage-driven ion migration in bipolar RRAM—Part II: Modeling. IEEE Transactions on Electron Devices. 2012;**59**:2468-2475. DOI: 10.1109/TED.2012.2202320

[21] Shang DS, Sun JR, Shi L, Wang ZH, Shen BG. Resistance dependence of photovoltaic effect in Au/$SrTiO_3$:Nb(0.5 wt%) Schottky junctions. Applied Physics Letters. 2008;**93**:172119. DOI: 10.1063/1.3009285

[22] Lee E, Gwon M, Kim D-W, Kim H. Resistance state-dependent barrier inhomogeneity and transport mechanisms in resistive-switching Pt/$SrTiO_3$ junctions. Applied Physics Letters. 2011;**98**:132905. DOI: 10.1063/1.3567755

[23] Strachan JP, Strukov DB, Borghetti J, Joshua Yang J, Medeiros-Ribeiro G, Stanley Williams R. The switching location of a bipolar memristor: Chemical, thermal and structural mapping. Nanotechnology. 2011;**22**:254015. DOI: 10.1088/0957-4484/22/25/254015

[24] Choi JS, Kim J-S, Hwang IR, Hong SH, Jeon SH, Kang S-O, et al. Different resistance switching behaviors of NiO thin films deposited on Pt and $SrRuO_3$ electrodes. Applied Physics Letters. 2009;**95**:022109. DOI: 10.1063/1.3173813

[25] Asamitsu A, Tomioka Y, Kuwahara H, Tokura Y. Current switching of resistive states in magnetoresistive manganites. Nature. 1997;**388**:50-52. DOI: 10.1038/40363

[26] Kim DS, Kim YH, Lee CE, Kim YT. Colossal electroresistance mechanism in a Au/$Pr0.7Ca0.3MnO_3$/Pt sandwich structure: Evidence for a Mott transition. Physical Review B. 2006;**74**:174430. DOI: 10.1103/PhysRevB.74.174430

[27] Meijer GI, Staub U, Janousch M, Johnson SL, Delley B, Neisius T. Valence states of Cr and the insulator-to-metal transition in Cr-doped $SrTiO_3$. Physical Review B. 2005;**72**:155102. DOI: 10.1103/PhysRevB.72.155102

[28] Yan ZB, Wang KF, Li SZ, Luo SJ, Liu J-M. Reversible resistance switching in $La0.225Pr0.4Ca0.375MnO_3$: The joule-heat-assisted phase transition. Applied Physics Letters. 2009;**95**:143502. DOI: 10.1063/1.3241997

[29] Fors R, Khartsev SI, Grishin AM. Giant resistance switching in metal-insulator-manganite junctions: Evidence for Mott transition. Physical Review B. 2005;**71**:045305. DOI: 10.1103/PhysRevB.71.045305

[30] Waser R, Aono M. Nanoionics-based resistive switching memories. Nature Materials. 2007;**6**:833-840. DOI: 10.1038/nmat2023

[31] Jiang W, Noman M, Lu YM, Bain JA, Salvador PA, Skowronski M. Mobility of oxygen vacancy in $SrTiO_3$ and its implications for oxygen-migration-based resistance switching. Journal of Applied Physics. 2011;**110**:034509. DOI: 10.1063/1.3622623

[32] Hong S, Choi T, Jeon JH, Kim Y, Lee H, Joo H-Y, et al. Large resistive switching in ferroelectric $BiFeO_3$ nano-island based switchable diodes. Advanced Materials. 2013;**25**:2339-2343. DOI: 10.1002/adma.201204839

[33] Valencia S, Crassous A, Bocher L, Garcia V, Moya X, Cherifi RO, et al. Interface-induced room-temperature multiferroicity in $BaTiO_3$. Nature Materials. 2011;**10**:753-758. DOI: 10.1038/nmat3098

[34] Green MA, Hishikawa Y, Dunlop ED, Levi DH, Hohl-Ebinger J, Ho-Baillie AWY. Solar cell efficiency tables (version 52). Progress in Photovoltaics: Research and Applications. 2018;**26**:427-436

[35] Unger EL, Hoke ET, Bailie CD, Nguyen WH, Bowring AR, Heumüller T, et al. Hysteresis and transient behavior in current-voltage measurements of hybrid-perovskite absorber solar cells. Energy & Environmental Science. 2014;**7**:3690-3698. DOI: 10.1039/C4EE02465F

[36] Snaith HJ, Abate A, Ball JM, Eperon GE, Leijtens T, Noel NK, et al. Anomalous hysteresis in perovskite solar cells. Journal of Physical Chemistry Letters. 2014;**5**:1511-1515. DOI: 10.1021/jz500113x

[37] Gu C, Lee JS. Flexible hybrid organic-inorganic perovskite memory. ACS Nano. 2016;**10**:5413-5418. DOI: 10.1021/acsnano.6b01643

[38] Yoo EJ, Lyu M, Yun J, Kang CJ, Choi YJ, Wang L. Resistive switching behavior in organic-inorganic hybrid $CH_3NH_3PbI_{3-x}Cl_x$ perovskite for resistive random access memory devices. Advanced Materials. 2015;**27**:6170-6175

[39] Hwang B, Gu C, Lee D, Lee J-S. Effect of halide-mixing on the switching behaviors of organic-inorganic hybrid perovskite memory. Scientific Reports. 2017;7:43794

[40] Senanayak SP, Yang B, Thomas TH, Giesbrecht N, Huang W, Gann E, et al. Understanding charge transport in lead iodide perovskite thin-film field-effect transistors. Science Advances. 2017;**3**:e1601935

[41] Wei W, Zhang Y, Xu Q, Wei H, Fang Y, Wang Q, et al. Monolithic integration of hybrid perovskite single crystals with heterogenous substrate for highly sensitive X-ray imaging. Nature Photonics. 2017;**11**:315

[42] Ward JW, Smith HL, Zeidell A, Diemer PJ, Baker SR, Lee H, et al. Solution-processed organic and halide perovskite transistors on hydrophobic surfaces. ACS Applied Materials & Interfaces. 2017;**9**:18120-18126

[43] Xu W, Cho H, Kim YH, Kim YT, Wolf C, Park CG, et al. Organometal halide perovskite artificial synapses. Advanced Materials. 2016;**28**:5916-5922. DOI: 10.1002/adma.201506363

[44] Choi J, Han JS, Hong K, Kim SY, Jang HW. Organic-inorganic hybrid halide perovskites for memories, transistors, and artificial synapses. Advanced Materials. 2018;**30**:1704002. DOI: 10.1002/adma.201704002

[45] Dang VQ, Han G-S, Trung TQ, Jin Y-U, Hwang B-U, Jung H-S, et al. Methylammonium lead iodide perovskite-graphene hybrid channels in flexible broadband phototransistors. Carbon. 2016;**105**:353-361

[46] Kim H, Han JS, Choi J, Kim SY, Jang HW. Halide perovskites for applications beyond photovoltaics. Small Methods. 2018;**2**:1700310

[47] Huo C, Liu X, Song X, Wang Z, Zeng H. Field-effect transistors based on van-der-Waals-grown and dry-transferred all-inorganic perovskite ultrathin platelets. Journal of Physical Chemistry Letters. 2017;**8**:4785-4792. DOI: 10.1021/acs.jpclett.7b02028

[48] Ma H, Wang W, Xu H, Wang Z, Tao Y, Chen P, et al. Interface state-induced negative differential resistance observed in hybrid perovskite resistive switching memory. ACS Applied Materials & Interfaces. 2018;**10**:21755-21763. DOI: 10.1021/acsami.8b07850

[49] Lee Y, Lee T-W. Organic synapses for neuromorphic electronics: From brain-inspired computing to sensorimotor nervetronics. Accounts of Chemical Research. 2019;**52**:964-974. DOI: 10.1021/acs.accounts.8b00553

[50] D'Innocenzo V, Kandada ARS, De Bastiani M, Gandini M, Petrozza A. Tuning the light emission properties by band gap engineering in hybrid lead halide perovskite. Journal of the American Chemical Society. 2014;**136**:17730-17733. DOI: 10.1021/ja511198f

[51] Calabrese J, Jones NL, Harlow RL, Herron N, Thorn DL, Wang Y. Preparation and characterization of layered lead halide compounds. Journal of the American Chemical Society. 1991;**113**:2328-2330

[52] Amat A, Mosconi E, Ronca E, Quarti C, Umari P, Nazeeruddin MK, et al. Cation-induced band-gap tuning in organohalide perovskites: Interplay of spin-orbit coupling and octahedra tilting. Nano Letters. 2014;**14**:3608-3616

[53] Koh TM, Krishnamoorthy T, Yantara N, Shi C, Leong WL, Boix PP, et al. Formamidinium tin-based perovskite with low Eg for photovoltaic applications. Journal of Materials Chemistry A. 2015;**3**:14996-15000

[54] Hao F, Stoumpos CC, Cao DH, Chang RPH, Kanatzidis MG. Lead-free solid-state organic-inorganic halide perovskite solar cells. Nature Photonics. 2014;**8**:489-494. DOI: 10.1038/nphoton.2014.82

[55] Huang C, Gao Y, Wang S, Zhang C, Yi N, Xiao S, et al. Giant blueshifts of excitonic resonances in two-dimensional lead halide perovskite. Nano Energy. 2017;**41**:320-326

[56] Zhang J, Morbidoni M, Huang K, Feng S, McLachlan MA. Environmentally friendly, aqueous processed ZnO as an efficient electron transport layer for low temperature processed metal-halide perovskite photovoltaics. Inorganic Chemistry Frontiers. 2018;**5**:84-89. DOI: 10.1039/C7QI00667E

[57] Zhang J, Tan CH, Du T, Morbidoni M, Lin C, Xu S, et al. ZnO-PCBM bilayers as electron transport layers in low-temperature processed perovskite solar cells. Scientific Bulletin. 2018;**63**:343-348. DOI: 10.1016/j.scib.2018.02.004

[58] Shao Y, Fang Y, Li T, Wang Q, Dong Q, Deng Y, et al. Grain boundary dominated ion migration in polycrystalline organic-inorganic halide perovskite films. Energy & Environmental Science. 2016;**501**:395-398. DOI: 10.1039/C6EE00413J

[59] Shao Y, Xiao Z, Bi C, Yuan Y, Huang J. Origin and elimination of photocurrent hysteresis by fullerene passivation in $CH_3NH_3PbI_3$ planar heterojunction solar cells. Nature Communications. 2014;**5**:1-7. DOI: 10.1038/ncomms6784

[60] Yang B, Brown CC, Huang J, Collins L, Sang X, Unocic RR, et al. Enhancing ion migration in grain boundaries of hybrid organic-inorganic perovskites by chlorine. Advanced Functional Materials. 2017;**27**:1700749

[61] Xiao Z, Huang J. Energy-efficient hybrid perovskite memristors and synaptic devices. Advanced Electronic Materials. 2016;**2**:1-8. DOI: 10.1002/aelm.201600100

[62] Wang Y, Lv Z, Chen J, Wang Z, Zhou Y, Zhou L, et al. Photonic synapses based on inorganic perovskite quantum dots for neuromorphic computing. Advanced Materials. 2018;**30**:1-9. DOI: 10.1002/adma.201802883

[63] Liu D, Lin Q, Zang Z, Wang M, Wangyang P, Tang X, et al. Flexible all-inorganic perovskite $CsPbBr_3$ nonvolatile memory device. ACS Applied Materials & Interfaces. 2017;**9**:6171-6176. DOI: 10.1021/acsami.6b15149

[64] Yan K, Chen B, Hu H, Chen S, Dong B, Gao X, et al. First fiber-shaped non-volatile memory device based on hybrid organic-inorganic perovskite. Advanced Electronic Materials. 2016;**2**:1-7. DOI: 10.1002/aelm.201600160

[65] Yu JC, Kim DB, Baek G, Lee BR, Jung ED, Lee S, et al. High-performance planar perovskite optoelectronic devices: A morphological and interfacial control by polar solvent treatment. Advanced Materials. 2015;**27**:3492-3500. DOI: 10.1002/adma.201500465

[66] Jeon NJ, Noh JH, Kim YC, Yang WS, Ryu S, Il Seok S. Solvent engineering for high-performance inorganic-organic hybrid perovskite solar cells. Nature Materials. 2014;**13**:1-7. DOI: 10.1038/nmat4014

[67] Lee MM, Teuscher J, Miyasaka T, Murakami TN, Snaith HJ. Efficient hybrid solar cells based on meso-superstructured organometal halide perovskites. Science. 2012;**338**:643-647. DOI: 10.1126/science.1228604

[68] Liu M, Johnston MB, Snaith HJ. Efficient planar heterojunction perovskite solar cells by vapour deposition. Nature. 2013;**501**:395-398. DOI: 10.1038/nature12509

[69] Liang P-W, Liao C-Y, Chueh C-C, Zuo F, Williams ST, Xin X-K, et al. Additive enhanced crystallization of solution-processed perovskite for highly efficient planar-heterojunction solar cells. Advanced Materials. 2014;**26**:3748-3754. DOI: 10.1002/adma.201400231

[70] Choi J, Park S, Lee J, Hong K, Kim D-H, Moon CW, et al. Organolead halide perovskites for low operating voltage multilevel resistive switching. Advanced Materials. 2016;**28**:6562-6567. DOI: 10.1002/adma.201600859

[71] Hwang B, Lee J. A strategy to design high-density nanoscale devices utilizing vapor deposition of metal halide perovskite materials. Advanced Materials. 2017;**29**:1701048

[72] Han JS, Van Le Q, Choi J, Kim H, Kim SG, Hong K, et al. Lead-free all-inorganic cesium tin iodide perovskite for filamentary and Interface-type resistive switching toward environment-friendly and temperature-tolerant nonvolatile memories. ACS Applied Materials & Interfaces. 2019;**11**:8155-8163. DOI: 10.1021/acsami.8b15769

[73] Zhou L, Chen X, Han S-T, Zhou Y, Lv Z, Wang Y, et al. Photonic synapses based on inorganic perovskite quantum dots for neuromorphic computing. Advanced Materials. 2018;**30**:1802883. DOI: 10.1002/adma.201802883

[74] Zhang H, Li Q, Song X, Zhang H, Zhang Y, Yu Y, et al. Low-voltage all-inorganic perovskite quantum

dot transistor memory. Applied Physics Letters. 2018;**112**:212101. DOI: 10.1063/1.5028474

[75] Wu Y, Wei Y, Huang Y, Cao F, Yu D, Li X, et al. Capping $CsPbBr_3$ with ZnO to improve performance and stability of perovskite memristors. Nano Research. 2017;**10**:1584-1594

[76] Tian H, Zhao L, Wang X, Yeh YW, Yao N, Rand BP, et al. Extremely low operating current resistive memory based on exfoliated 2D perovskite single crystals for neuromorphic computing. ACS Nano. 2017;**11**:12247-12256. DOI: 10.1021/acsnano.7b05726

[77] Kang K, Ahn H, Song Y, Lee W, Kim J, Kim Y, et al. High-performance solution-processed organo-metal halide perovskite unipolar resistive memory devices in a cross-bar array structure. Advanced Materials. 2019;**31**:e1804841. DOI: 10.1002/adma.201804841

[78] Cheng X-F, Hou X, Zhou J, Gao B-J, He J-H, Li H, et al. Pseudohalide-induced 2D $(CH_3NH_3)_2PbI_2(SCN)_2$ perovskite for ternary resistive memory with high performance. Small. 2018;**14**:1703667. DOI: 10.1002/smll.201703667

[79] Kumar M, Kim H-S, Park DY, Jeong MS, Kim J. Compliance-free multileveled resistive switching in a transparent 2D perovskite for neuromorphic computing. ACS Applied Materials & Interfaces. 2018;**10**:12768-12772. DOI: 10.1021/acsami.7b19406

[80] Hu Y, Zhang S, Miao X, Su L, Bai F, Qiu T, et al. Ultrathin Cs3Bi2I9 nanosheets as an electronic memory material for flexible memristors. Advanced Materials Interfaces. 2017;**4**:2-9. DOI: 10.1002/admi.201700131

[81] Chen J-Y, Chiu Y-C, Li Y-T, Chueh C-C, Chen W-C. Nonvolatile perovskite-based photomemory with a multilevel memory behavior. Advanced Materials. 2017;**29**:1702217. DOI: 10.1002/adma.201702217

[82] Wang Y, Lv Z, Liao Q, Shan H, Chen J, Zhou Y, et al. Synergies of electrochemical metallization and valance change in all-inorganic perovskite quantum dots for resistive switching. Advanced Materials. 2018;**30**:e1800327. DOI: 10.1002/adma.201800327

[83] Tian H, Wang X, Wu F, Yang Y, Ren T-L. High performance 2D perovskite/graphene optical synapses as artificial eyes. In: 2018 IEEE Int. Electron Devices Meeting; IEEE. 2018. pp. 38.6.1-38.6.4

[84] Qian L, Sun Y, Wu M, Li C, Xie D, Ding L, et al. A lead-free two-dimensional perovskite for a high-performance flexible photoconductor and a light-stimulated synaptic device. Nanoscale. 2018;**10**:6837-6843. DOI: 10.1039/c8nr00914g

Chapter 9

Osmium Containing Double Perovskite Ba_2XOsO_6 (X = Mg, Zn, Cd): Important Candidates for Half-Metallic Ferromagnetic and Spintronic Applications

Sajad Ahmad Dar

Abstract

In this chapter, Osmium-based double perovskites Ba_2XOsO_6(X = Mg, Zn, Cd) have been investigated for their magnetic structure, electronic, elastic, mechanical and thermodynamic belongings. These materials have been recently reported experimentally for their magnetic structure. Here, we report the first successful ab initio calculations on the physical properties of these materials. The structural optimization for these Ba_2XOsO_6(X = Mg, Zn, Cd) double perovskite compounds has been finalized within density functional theory via full potential linearized augmented plane wave (FP-LAPW) method. The structural investigation exposes the ferromagnetic phase stability of these compounds. The spin-polarized electronic and magnetic properties were calculated within generalized gradient approximation (GGA), Hubbard approximation (GGA+U) and modified Becke-Johnson approximation (mBJ). The electronic profile establishes the half-metallic nature for all the three compounds. The total spin magnetic moment was found to be an integer value of 2 μ_b. The elastic constants have been calculated and used to predict mechanical stuffs like Shear modulus (G), Poisson ratio (v) and anisotropic factor. The calculated B/G and Cauchy pressure (C_{12}-C_{44}) both characterize these materials as brittle. The thermodynamic parameters like heat capacity and Debye temperature have been predicted in the temperature range of 0–1000 K.

Keywords: Ba_2XOsO_6 (X = Mg, Zn, Cd), spintronics, ferromagnetic, elastic, mechanical behavior, thermodynamics

1. Introduction

The need of advanced materials with novel properties for industrial and technological use has strained the material community to have a deep and appropriate understanding of the periodic table elements, along with their combinations. Therefore, materials community consequently observes the vital changes in innovative designing of novel materials. A tremendous increase in simulation power, along with algorithmic improvements in quantum theory allows one to have well-organized and exact quantum mechanical calculations. This has hence stretched the computing power to such extent that those properties of materials which were once observed extremely difficult are now easily being calculated with a great precision [1, 2]. From last few years work on perovskites especially double perovskites has geared up due to their vast technological applications and displaying multifunctional properties. The general formula of perovskites is looked as ABO_3, where "A" and "B" are cations and "O" is oxygen anion. The charge of "A" and "B" cations can vary in the original Perovskite. Double perovskites are potential members of this diverse perovskite family having different structures, composition and properties. The double perovskite compounds having a general formula $A_2BB'O_6$ have benefited the material community because of great technological applications including spintronic materials, multi-ferroic materials, half-metallic materials, ferromagnetic materials, magneto-dielectric materials, magneto-optic materials, insulating ferrimagnetism [3–8]. The double perovskite family exhibits a wide range of magnetic behaviors, like actually simple antiferromagnets presented by (Ba_2LiOsO_6), ferromagnets (Ba_2MgReO_6 and Ba_2NaOsO_6), spin singlet ground states for (Ba_2YMoO_6), and spin glasses(Sr_2CaReO_6) [9–13]. Osmium based double perovskites especially Sr_2FeOsO_6, $SrFeCaOsO_6$, Ca_2FeOsO_6 have been extensively investigated for puzzling magnetic behavior [14–19]. Perovskites with ABO_3 structure like $BaPuO_3$, $SrPuO_3$, $BaAmO_3$, $SrAmO_3$, $EuGaO_3$, $EuInO_3$ etc. have been reported for spintronic applications [20–24]. Further materials like $BaMoO_3$, $SrMoO_3$, $XReO_3$ (X = Rb, Cs, Tl), $SnTaO_3$, $PbMoO_3$ [25–28] have been recently reported for fuel cell applications. Numerous investigations have also been reported on halide perovskites for solar cell applications like $MAPbBr_3$ and $MAPbI_3$ [29–36].

Ba_2XOsO_6(X = Mg, Zn, Cd) cubic double perovskites have been recently reported in space group Fm-3 m (225). The complete details of the respective experimental lattice parameters are given in **Table 1**. The Ba atoms are located at 8c (1/4, 1/4, 1/4) of the unit cell, X atoms are at position 4b (0.5, 0.5, 0.5), Os atoms are positioned at 41a (0, 0, 0) and O atoms at 24e (X, 0, 0) (X = 0.233, 0.238, 0.234)

Parameter	Present	Other	Present	Other	Present	Other
	Ba_2MgOsO_6		Ba_2ZnOsO_6		Ba_2CdOsO_6	
Lattice Constant	8.1548	8.07 [37] 8.08 [61] 8.06 [61]	8.19	8.09 [37] 8.09 [61] 8.06 [61]	8.388	8.31 [37] 8.325 [61]
Volume	914.90		929.62		995.96	
B	150.93		144.23		140.14	
B'	4.44		4.13		4.7	
Bond length						
Os-O	1.95		1.95		2.00	
Mg, Zn, Cd-O	2.11		2.15		2.18	
Ba-Ba	4.07		4.10		4.19	
Ba, Mg, Zn, Cd	3.53		3.55		3.62	
Os-Mg, Zn, Cd	4.07		4.10		4.19	
E_0	−68425.9		−71617.3		−79217.0	

Table 1.
Optimized ground states parameters for all the three osmium double perovskites.

respectively for Mg, Zn, Cd [37]. Further double perovskites of the variant $A_2BB'O_6$ like Ba_2MgReO_6, Sr_2MnTaO_6, Ba_2InTaO_6 and many more have also been reported for electronic, magnetic, mechanical, optical, thermoelectric and thermodynamic investigations [38–47]. As far Ba_2XOsO_6(X = Mg, Zn, Cd) compounds are concerned which belongs to the same variant not much attention has been paid towards these perovskites so far, regarding the above mentioned characteristic properties. Hence, in the present work an attempt to predict the properties for these double perovskite has been made and to check out their potential applications. The most successful density functional theory (DFT) has been employed for the investigation of magnetic, electronic, elastic, mechanical and thermo-physical behavior. For the investigation of thermo-physical behavior quasi harmonic Debye model [48, 49] has been used for the prediction of important parameters like specific heat, thermal expansion, Debye temperature, Grüneisen parameter etc.

2. Computational details

The computational technique used during the calculations process is based on full-potential linearized augmented plane wave (FP-LAPW) [50] method based upon density functional theory (DFT) [51] as employed in WIEN2K. For structural optimization generalized gradient approximation (GGA) scheme of Perdew, Burke and Ernzerhof (PBE) [52] has been used. For electronic and magnetic calculations in addition to (GGA), Hubbard approximation (GGA + U) [45] and modified Becke-Johnson (mBJ) [53] has been used. For GGA + U approach the incorporation U- term can be done by various methods [54, 55]. In the present work we have used self-interaction correction method (SIC) [56] as implemented in WIEN2K. The value of U_{eff} was varied from 1 to 5 eV and J was set to 0, so as to properly adjust the Os-d in density of states. The final U value used throughout the calculations was set to 2.00 eV [57]. For precise energy convergence the value of $R_{MT}K_{max}$ was taken 7, where R_{MT} is the small atomic radius in unit cell and K_{max} denotes the size of the largest **k** vector in the plane wave expansion. The value of L_{max} was taken as 10, and G_{max} = 12 $(a.u.)^{-1}$. The energy and charge convergence criterion is considered when the total energy is stable within 0.001 Ry and the charge difference is less than $0.001e/a.u.^3$ per unit cell. A mesh of 1000 K points is considered for Brillouin zone integration via tetrahedral method [58]. The elastic constants were calculated using the scheme developed by Charpin [59] as integrated in WIEN2K package. The thermodynamic parameters have been calculated using quasi-harmonic Debye model [48, 49] for the pressure and temperature dependency of some essential thermodynamic parameters. In this model the Gibbs function takes the form;

$$G*(V,P,T) = E(V) + P(V) + F_{vib}\,(\theta(V);\,T) \tag{1}$$

where E(V), P(V), $\theta(V)$ are the total energy per unit cell, the constant hydrostatic pressure and Debye temperature respectively and F_{vib} is the vibration term written as;

$$F_{vib}\,[\theta(V);\,T] = N\,K_B T\left[\frac{9\theta}{8T} + 3\ln\left(1 - e^{\frac{-\theta}{T}}\right) - D\left(\frac{\theta}{T}\right)\right] \tag{2}$$

$D\left(\frac{\theta}{T}\right)$ in equation represents the Debye integral, N is the number of atoms per formula unit, K_B the Boltzmann's constant. The Debye temperature θ_D is expressed as;

$$\theta_D = \frac{h}{K_B}\left(6\pi^2 V^{1/2} N\right)^{1/3} f(v)\sqrt{\frac{B_s}{M}} \tag{3}$$

In the above equation B_s represents the adiabatic bulk modulus, M is the molecular mass per unit cell, the bulk modulus is expressed by

$$B_s = V\left(\frac{d^2E(V)}{dV^2}\right) \tag{4}$$

The non-equilibrium Gibbs function G*(V, P, T) can be minimized with respect to volume V;

$$\left[\frac{dG*(V;P,T)}{dV}\right]_{P,T} \tag{5}$$

Solution of Eq. (5) gives a detailed information about the thermodynamic quantities like thermal expansion α, heat capacity at constant volume C_V, heat capacity at constant pressure C_P, given respectively by;

$$\alpha = \frac{\gamma C_V}{B_T V} \tag{6}$$

$$C_V = 3nk\left[4D\left(\frac{\theta_D}{T}\right) - \frac{3\theta_D/T}{e^{\frac{\theta_D}{T}} - 1}\right] \tag{7}$$

$$C_P = C_V(1 + \gamma\alpha T) \tag{8}$$

In Eq. (8) γ represents the Grüneisen parameter, which is approximated as

$$\gamma = -\frac{dln\theta_D(V)}{dln\ V} \tag{9}$$

3. Results and discussion

3.1 Structural properties

The optimized volume for all the three compounds has been made by fitting the total energy as a function of its cell volume using Birch–Murnaghan's equation of state [60]. Marjerrison et al. [37] have recently reported all the three compounds in cubic B1-phase space group Fm-3 m (225). The Ba atoms are located at position 8c (0.25, 0.25, 0.25), Mg, Zn, Cd at 4b (0.5, 0.5, 0.5), Os at 4a (0, 0, 0) and O atoms are sited at24e (X, 0, 0) (X = 0.233, 0.238, 0.234) respectively for Mg, Zn, Cd. The geometry and structural optimization has been carried in Non-magnetic (NM), ferromagnetic (FM), and anti-ferromagnetic (AFM) phases. The ground state energy was found lowest for all the three compounds in the ferromagnetic phase as presented in **Figure 1(a–c)**, and thus a stable configuration.

The optimized ground state lattice constants are close to available experimental and theoretical results. The ground state parameters like bulk modulus (B_0), lattice constant (a_0) pressure derivatives of bulks modulus and energy are grouped in **Table 1**.

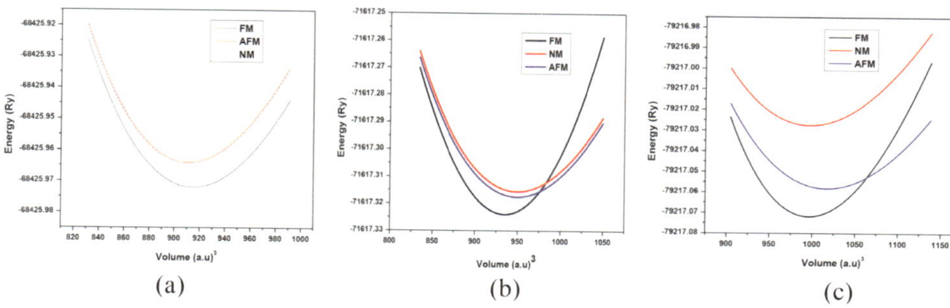

Figure 1.
Energy versus volume for ferromagnetic (FM), non-magnetic (NM) and anti-ferromagnetic (AFM) cases (a) Ba_2MgOsO_6 (b) Ba_2ZnOsO_6 (c) Ba_2CdOsO_6.

3.2 Elastic and mechanical properties

In the present work the Charpin method has been employed for the calculation of elastic constants C_{ij} (C_{11}, C_{12}, C_{44}) values as implemented in WIEN2K. The values of elastic constants were obtained by calculating the total energy as a function of volume-conserving strains. The value of the elastic constants and mechanical properties are summed up in **Table 2**. The calculated values of elastic constants properly satisfy the criteria for cubic elastic constants and ensures the stability $C_{11}–C_{12} > 0$, $C_{11} > 0$, $C_{44} > 0$, $(C_{11} + 2C_{12}) > 0$, $C_{12} < B < C_{11}$ [47]. The Poisson's ratio (ν), Young's modulus (E), and Shear modulus (G) are calculated by using [62–64] and presented in **Table 2**. According to Hill [65] average shear modulus, *G* is defined as arithmetic mean of Voigt, G_V and Reuss, G_R values. Young's modulus (E) deals with the stiffness of the material. The obtained value of (E) was found to be 215.75, 190.82, 169.83 GPa respectively for Ba_2XOsO_6(X = Mg, Zn, Cd). Thus large value of (E) provides a clear indication that these compounds will behave as tough materials. Ba_2MgOsO_6 has the largest value of (E) as compared to other

GGA	$BaMgOsO_6$	$BaZnOsO_6$	Ba_2CdOsO_6
C_{11}	262.60	240.70	232.07
C_{12}	86.90	98.85	84.32
C_{44}	84.10	76.91	61.66
B	150	145.59	132.95
G_V	85.61	74.51	66.54
G_R	85.57	74.39	66.02
G	85.50	74.45	66.28
E	215.75	190.82	169.83
ν	0.2617	0.2815	0.2871
B/G	1.7652	1.95	2.00
$C_{12}–C_{44}$	2.8	21.94	22.66
A	0.957	1.08	0.8346
T_m	2105 ± 300	2100 ± 300	1925 ± 300

Table 2.
Calculated elastic constants C_{11}, C_{12}, C_{44} in (GPa), bulk modulus B (GPa),shear modulus G (GPa), Young's modulus E (GPa), Poisson's ratio ν, Zener anisotropy factor a, B/G ratio, Cauchy pressure $C_{12}–C_{44}$ and melting temperature T_m (K) for Ba_2XOsO_6 (X = Mg, Zn, Cd).

perovskites under consideration in this study. The reason for the decreasing value of (E) is the Bulk modulus which has also a decreasing trend as one goes with X position from Mg to Cd via Zn. The B/G ratio is the measure of ductility and brittleness of a material. According to Pugh [66], a material is brittle if the ratio $B/G < 1.75$ and is ductile if $B/G > 1.75$. The B/G ratio for Ba_2XOsO_6(X = Mg, Zn, Cd), was calculated to be 1.765, 1.95, 2.00 respectively, which is higher than the limit value for all the three compounds, thus all the three compounds will show ductile nature. Cauchy pressure (C_{12}–C_{44}) also helps to estimate the ductility and brittleness of a material. The positive value of (C_{12}–C_{44}) portrays a material as ductile and negative value as brittle. The calculated value was also found to be positive for all the three compounds. Hence both B/G value and Cauchy pressure verifies the ductile nature for all the three perovskites Ba_2XOsO_6(X = Mg, Zn, Cd).

Zener anisotropy factor 'A' is the property of a material to show altered characteristic in various direction of its structure. As per this a material is isotropic if and only if 'A' factor has unit value or otherwise anisotropic. The calculated value of 'A' for the compound was found to 0.975, 1.08, 0.83 which is less than unity for Ba_2MgOsO_6 and Ba_2CdOsO_6 and greater than unity for Ba_2ZnOsO_6, hence in all the three cases deviating from unity, thus the materials will present anisotropic nature. Poisson's ratio (ν) describes the nature of bonding forces. The upper and lower limits of Poisson's ratio are 0.25 and 0.50 [20–24]. The (ν) value varies from material to material. For covalent materials, (ν) has a typical value of 0.1, for ionic materials (ν) = 0.25 and for metallic materials the value (ν) =0.33. The value of Poisson's ratio for Ba_2XOsO_6(X = Mg, Zn, Cd) was calculated to be 0.261, 0.281 and 0.287respectively, which lies close to 0.25 and hence suggest a higher ionic behavior as inter-atomic bonding for these compounds. The obtained values of elastic constants have also been used to predict, one important thermodynamic parameter known as melting temperature [26–28]. The calculated value of melting temperature was found 2100 ± 300, 2105 ± 300, 1925 ± 300 K respectively for Ba_2XOsO_6(X = Mg, Zn, Cd). The calculated values of elastic constants, mechanical properties including melting temperature are grouped in **Table 2**.

3.3 Electronic and magnetic properties

For electronic structure calculations spin resolved band structure and density of states have been plotted using different correlation potentials. GGA calculated lattice parameter has been used to plot band structure and density of states within GGA, GGA + U and mBJ. These band structures and density plots usually deliver a decent understanding of the electronic contour of a material. The combination of different methods for band structure and density of state plots has been done as to understand the variation of results within different correlations. **Figures 2(a–c)**, **3(a–c)** and **4(a–c)** represent the spin included band structures within GGA, GGA + U and mBJ respectively for Ba_2XOsO_6(X = Mg, Zn, Cd).

It is clear from these figures that the band profile for all the three compounds at the Fermi level is almost similar for all the approximation, presenting 100% of spin polarization. The Fermi level is set at 0 eV, separating the valance band maximum (VBM) from the conduction band minimum (CBM) in all figures. For spin up states the Fermi level remains fully occupied presenting metallic nature for all the three compounds and for spin down states the Fermi level remains completely vacant falling in a gap and thus generating a gap between(VBM) and (CBM), presenting the semi-conducting nature for the compounds. In case of Ba_2MgOsO_6 within GGA, GGA + U and mBJ respectively, the (VBM) lies on symmetry points 'Γ' at −1.30 eV, −1.1 eV and −1.3 eV, and CBM lies on symmetry point 'X' at 0.001, 0.7 and 1.2 eV respectively within GGA, GGA + U and mBJ. Hence in all the three cases the (VBM)

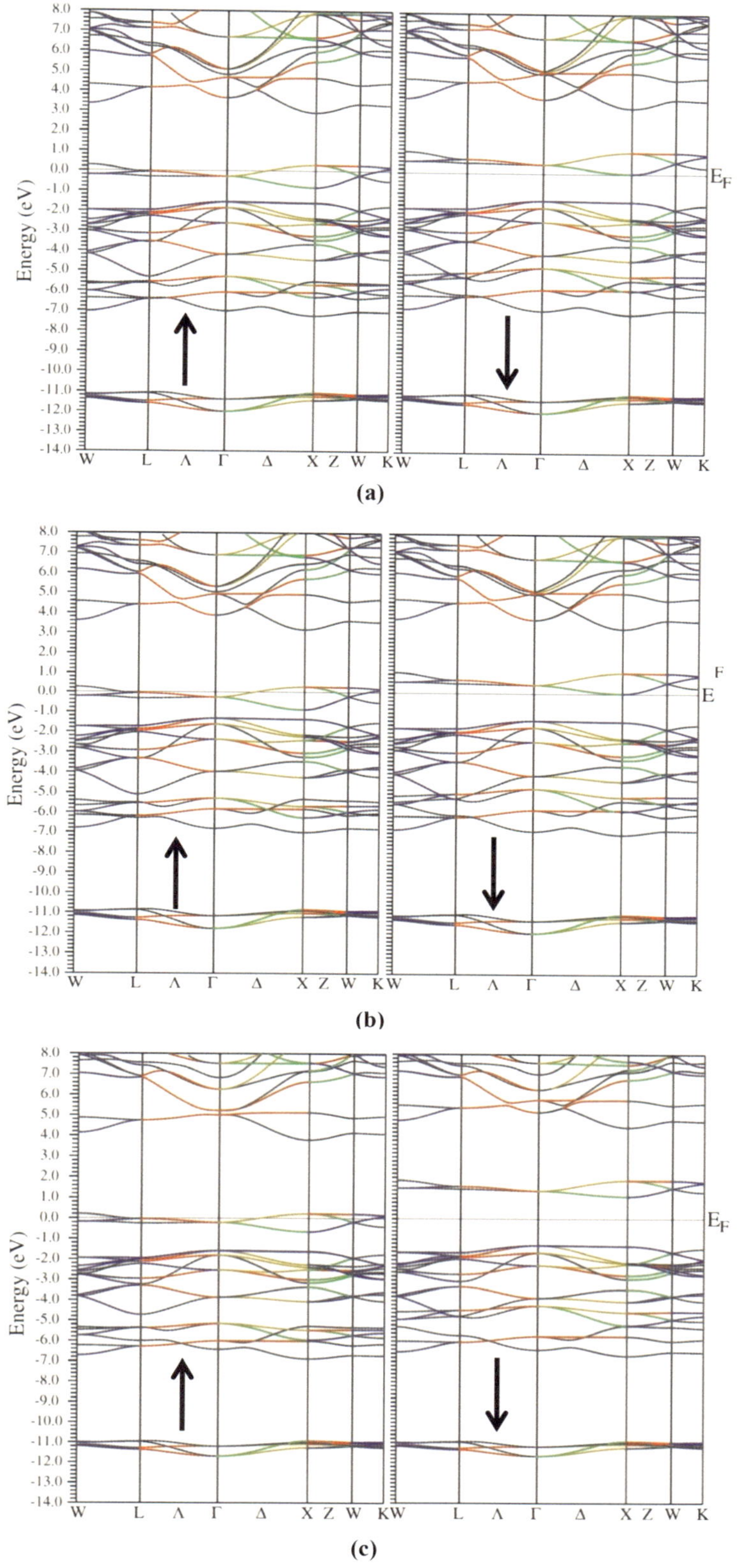

Figure 2.
Band structure left spin up and right spin down states within (a) GGA (b) GGA + U (c) mBJ for Ba_2MgOsO_6.

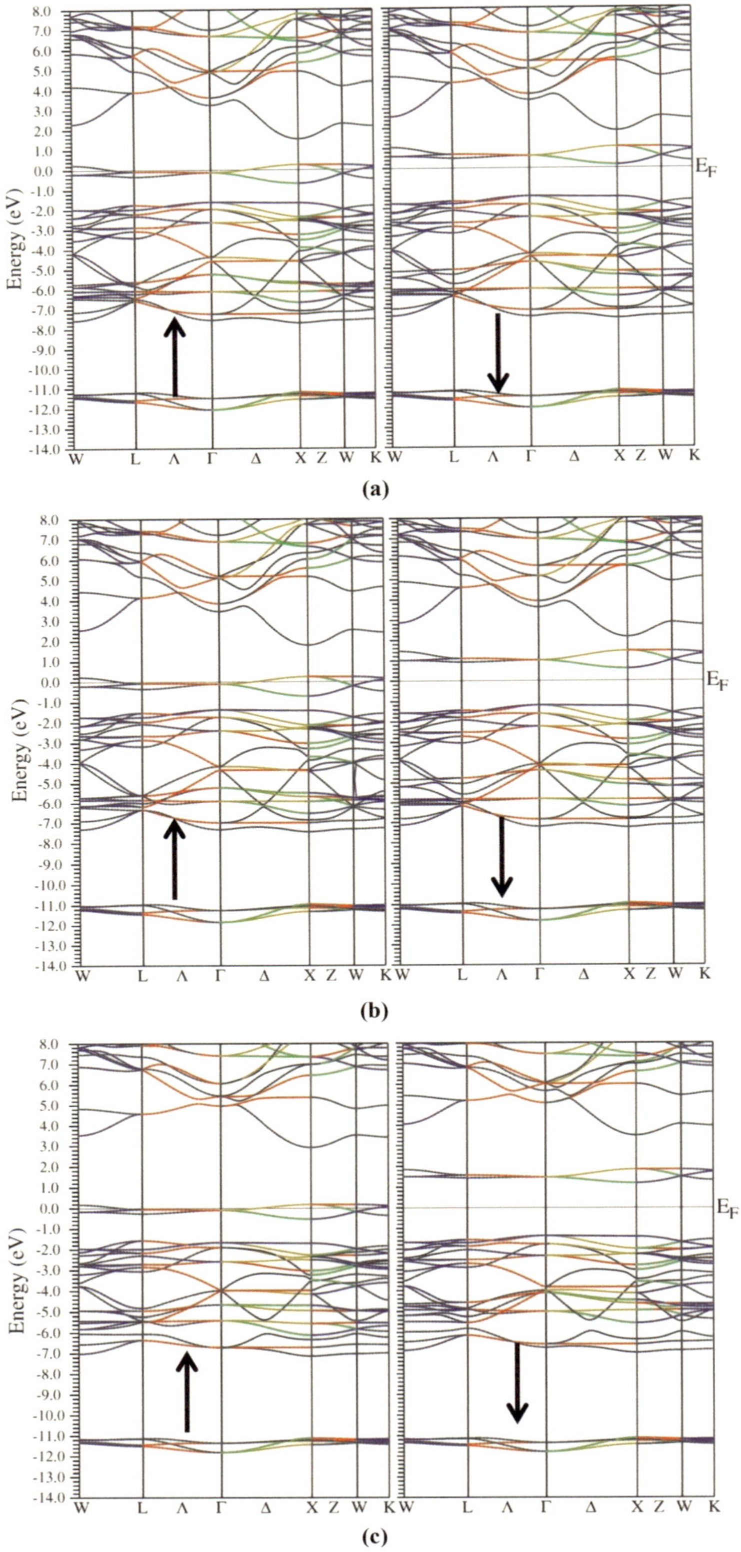

Figure 3.
Band structure left spin up and right spin down states within (a) GGA (b) GGA + U (c) mBJ for Ba_2ZnOsO_6.

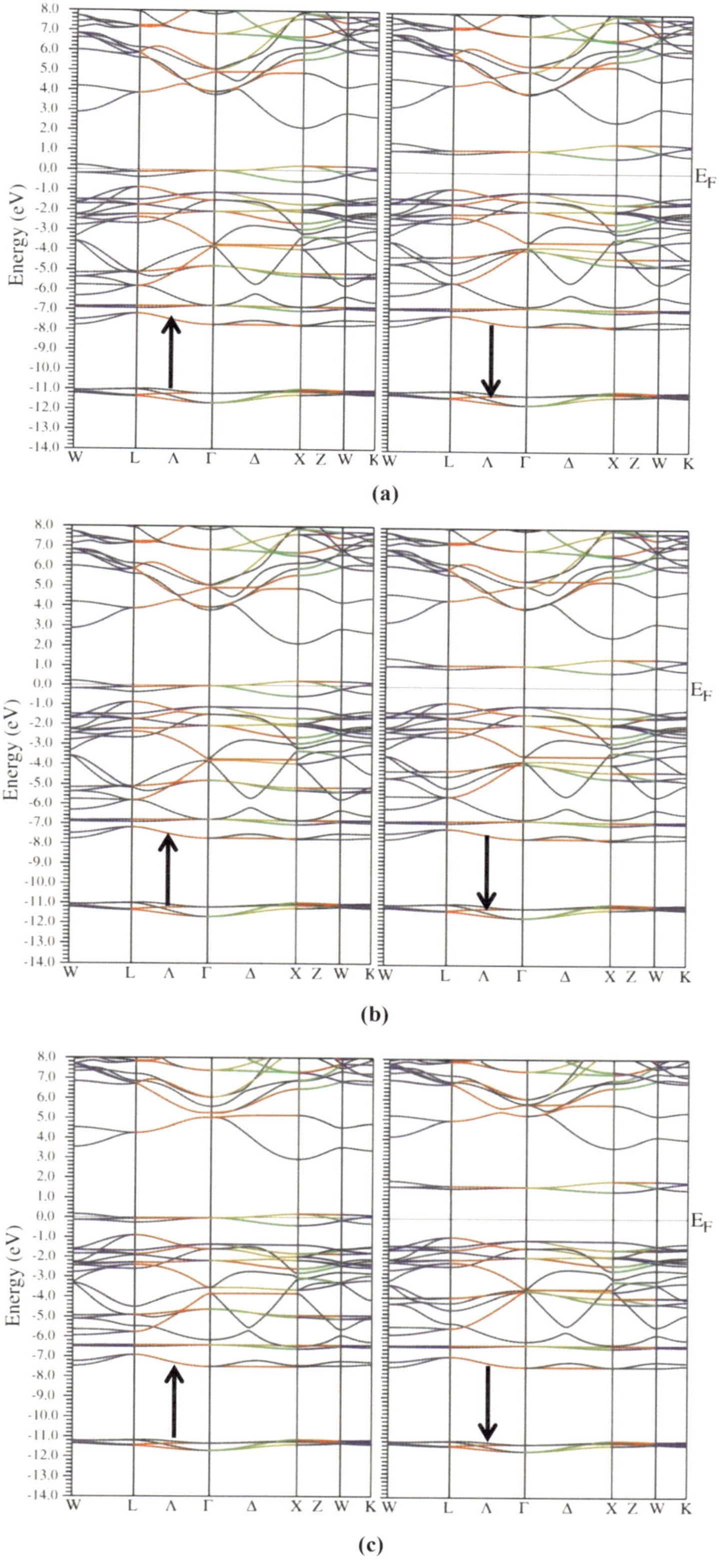

Figure 4.
Band structure of left spin up and right spin down (a) GGA (b) GGA + U (c) mBJ for Ba_2CdOsO_6.

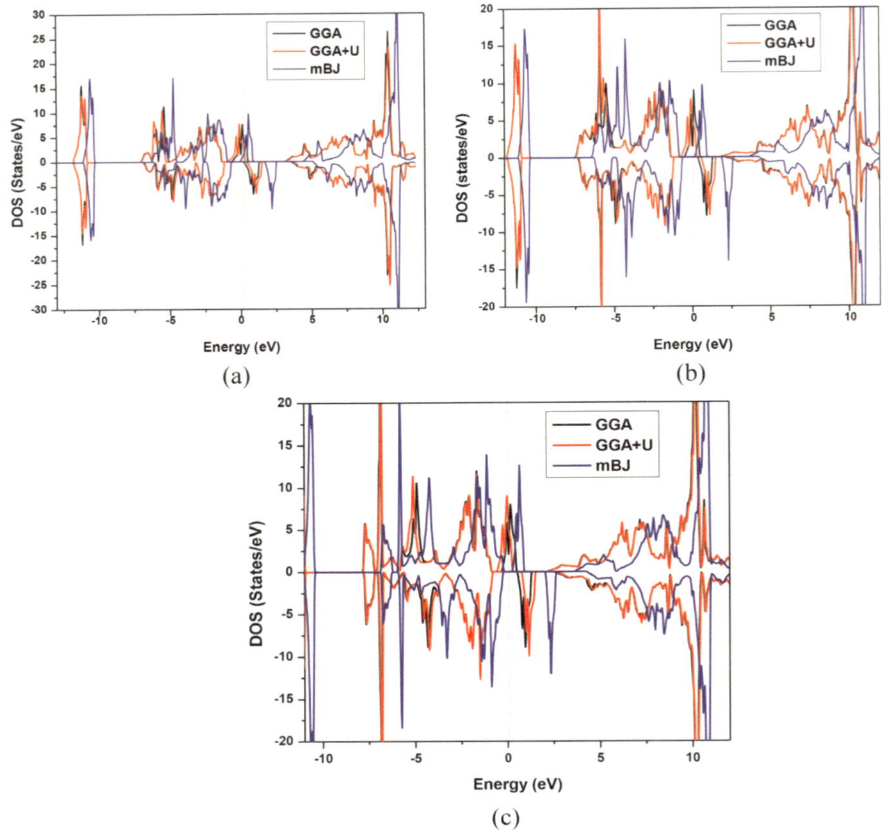

Figure 5.
Combined DOS diagram for spin up and down states within GGA, GGA + U and mBJ (a) Ba_2MgOsO_6 (b) Ba_2ZnOsO_6 (c) Ba_2CdOsO_6.

and (CBM) lie on 'Γ' and X point making the compound indirect band gap semi-conductor in spin down states. The band gap value changes as we apply U and mBJ, the value of band gap found in GGA, GGA + U and mBJ are 1.3, 1.8 and 2.5 eV respectively for Ba_2MgOsO_6. For Ba_2ZnOsO_6 the valance band maxima (VBM) lie on symmetry points 'Γ' at −1.40, −1.2, −1.4 eV respectively in GGA, GGA + U and mBJ, and the conduction band minimum (CBM) lies on symmetry point 'X' at 0.2, 0.7, and 1.0 eV respectively in GGA, GGA + U and mBJ, thus generating an indirect band gap of 1.6, 1.9 and 2.4 eV respectively for GGA, GGA + U and mBJ. Similarly for Ba_2CdOsO_6 the valance band maxima (VBM) lie on symmetry points 'L' at −1.00, −0.8, −0.9 eV respectively in GGA, GGA + U and mBJ, the conduction band minimum (CBM) lies on symmetry point 'X' at 0.5, 0.8, and 1.4 eV respectively in GGA, GGA + U and mBJ, thus generating an indirect band gap of 1.5, 1.6 and 2.3 eV respectively for GGA, GGA + U and mBJ. Thus from the band structure calculations 100% of spin polarization at Fermi level is observed. The compounds behave as metallic for spin up states and semi-conducting for spin down states. The overall band picture presents half-metallic nature for all the three compounds.

For the further explanation of the band picture, total density of states (TDOS) and partial density of states (PDOS) have been plotted. The spin included combined TDOS shown in **Figure 5(a–c)** disclose the same results as presented by band structure plots presenting metallic nature for spin up states and semi-conducting for spin down states for all the three approximations and hence overall half-metallic nature. The DOS peaks are found to increase in case of GGA + U and mBJ.

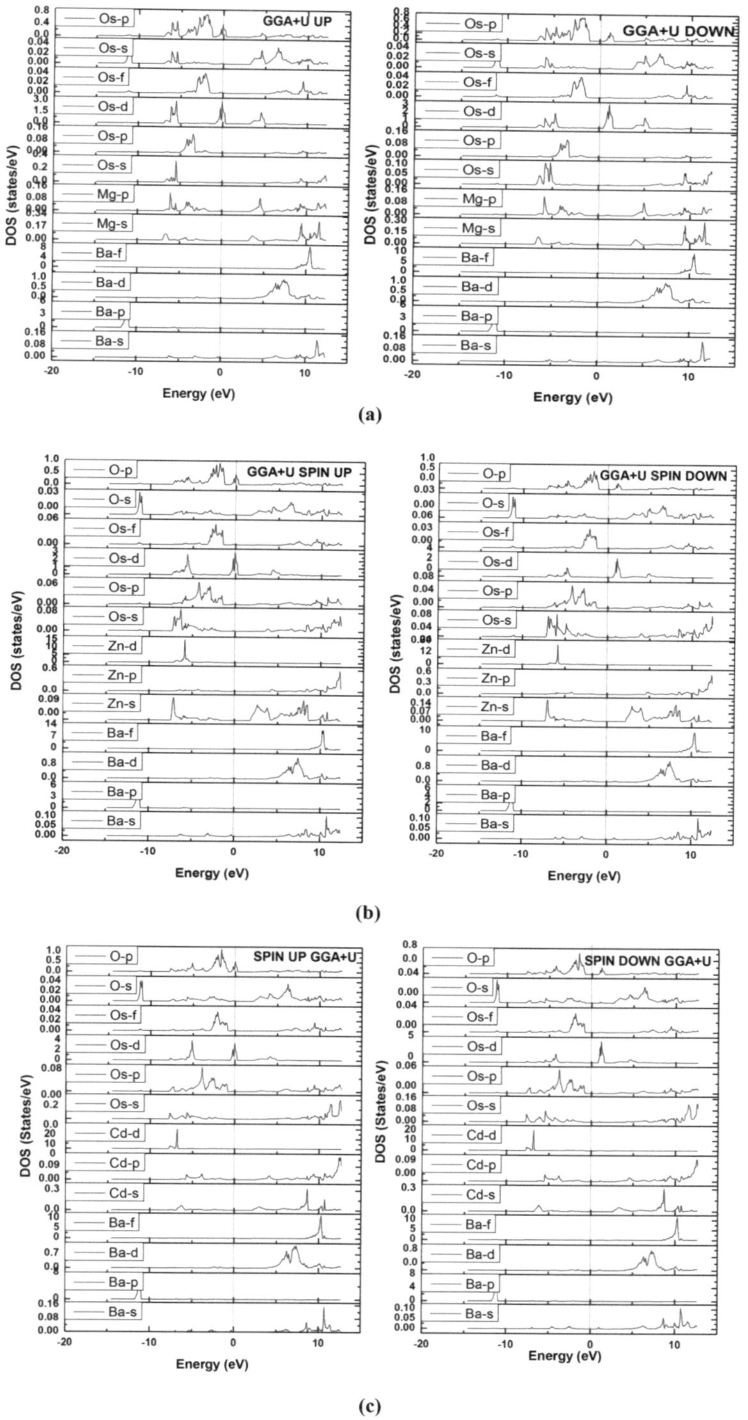

Figure 6.
Partial density of states contribution in spin up and down states within GGA + U (a) Ba_2MgOsO_6 (b) Ba_2ZnOsO_6 (c) Ba_2CdOsO_6.

The contribution to the TDOS picture has been represented by the partial contribution to the DOS diagram as depicted in **Figure 6(a–c)** for both spin up and down states within GGA + U. The (PDOS) has been plotted for (Ba-'s', 'p', 'd', 'f'),

(Mg- 's', 'p', Zn-'s', 'p', 'd'), (Cd-'s', 'p', 'd'), (Os- 's', 'p', 'd', 'f') and O-'s', 'p' states. From these figures. it is clear that the metallic nature in spin up states for the compound in all the approximations is due to the Os-'d' states which are present at Fermi level with a small contribution of O-'p' states hybridized with one another and in case of spin down these 'd'- states of Os and 'p'-states of O are pulled inside the conduction band, thereby generating a gap in spin down states. Thus the spin included band profile, TDOS and PDOS results display that Ba_2XOsO_6(X = Mg, Zn, Cd) all present half-metallic nature.

In order to check the magnetic nature of the compounds the total and partial magnetic moments have been calculated with GGA, GGA + U and mBJ. The total magnetic contribution is obtained as the summation of the partial moments of individual elements and the interstitial moments. The total magnetic moment obtained in all approximations is nearly same for all the compounds Ba_2XOsO_6(X = Mg, Zn, Cd) equal to an integer value 2 μ_B shown in **Table 3**.The main contribution to the total magnetic moment is mostly found from Osmium atoms. The partial moment of Os element shows a great variation on the application of Hubbard U and mBJ potentials. Hence it is clear that the ferromagnetic nature and large value of total magnetic moment for Ba_2XOsO_6(X = Mg, Zn, Cd) is mainly due to Os atoms. The values of interstitial, partial and total magnetic moments are present in **Table 3**. Thus the large and integer value of magnetic moment of 2 μ_B further verifies the half-metallic and ferromagnetic nature for Ba_2MgOsO_6. The integer value of magnetic moment is one of the criteria for the half-metallic nature of a compound [22, 23].

3.4 Thermodynamic properties

In order to check the thermodynamic behavior quasi-harmonic Debye approximation [26–28] has been employed to check the temperature and pressure variation of some noteworthy thermodynamic quantities like heat at constant volume (C_v), thermal expansion (α), Grüneisen parameter (γ), Debye temperature (θ_D) and also the bulk modulus variation for these double perovskites. The variation of these parameters has been investigated under pressure and temperature. The temperature has been varied from 0 to 1000 K and pressure ranges from 0 to 15 GPa, with the step size of 5 GPa pressure. In this range of temperature quasi harmonic Debye model remains unconditionally valid.

Figure 7(a–c) presents the variation of bulk modulus (B) with temperature at different pressure points respectively at 0, 5, 10, and 15 GPa. Our results present a clear decrease in bulk modulus with temperature and an increase is observed with

Compound	Method	M_{int}	M_{Ba}	$M/_{Mg,Zn,Cd}$	M_{Os}	M_O	M_{Tot}
Ba_2MgOsO_6	GGA	0.38	0.01	0.00	1.14	0.07	2.0
	GGA + U	0.35	0.01	0.00	1.25	0.06	2.0
	mBJ	0.22	0.00	0.00	1.41	0.05	2.0
Ba_2ZnOsO_6	GGA	0.37	0.01	0.00	1.15	0.07	2.0
	GGA + U	0.34	0.01	0.00	1.26	0.06	2.0
	mBJ	0.21	0.00	0.00	1.42	0.05	2.0
Ba_2CdOsO_6	GGA	0.37	0.01	0.00	1.18	0.06	2.0
	GGA + U	0.33	0.00	0.00	1.31	0.05	2.0
	mBJ	0.20	0.00	0.00	1.49	0.04	2.0

Table 3.
Calculated magnetic moment for Ferro-magnetic Ba_2XOsO_6(X = Mg, Zn, Cd). Within GGA, GGA + U and mBJ (in Bohr magneton μ_B).

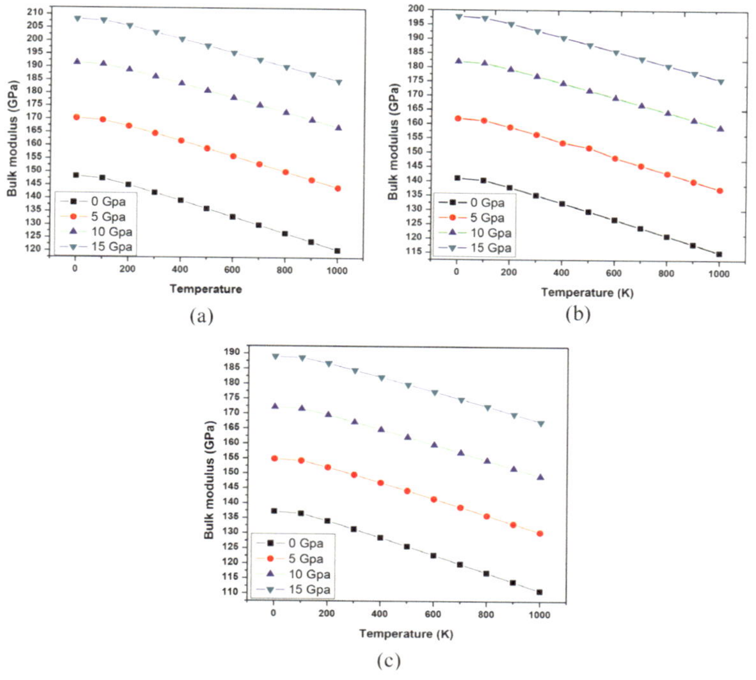

Figure 7.
Dependence of bulk modulus on temperature and pressures for (a) Ba_2MgOsO_6 (b) Ba_2ZnOsO_6 (c) Ba_2CdOsO_6.

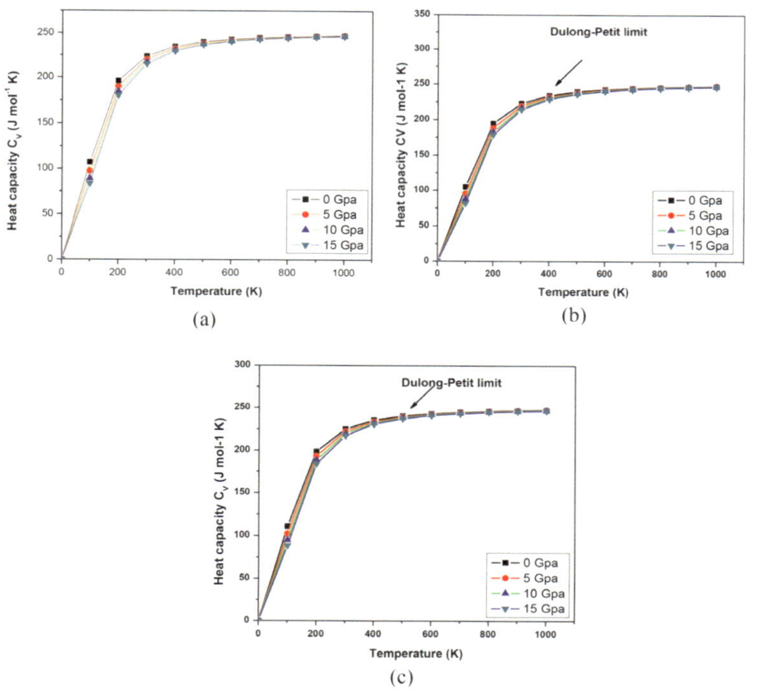

Figure 8.
Dependence of specific heat (C_V) on temperature and pressures for (a) Ba_2MgOsO_6 (b) Ba_2ZnOsO_6 (c) Ba_2CdOsO_6.

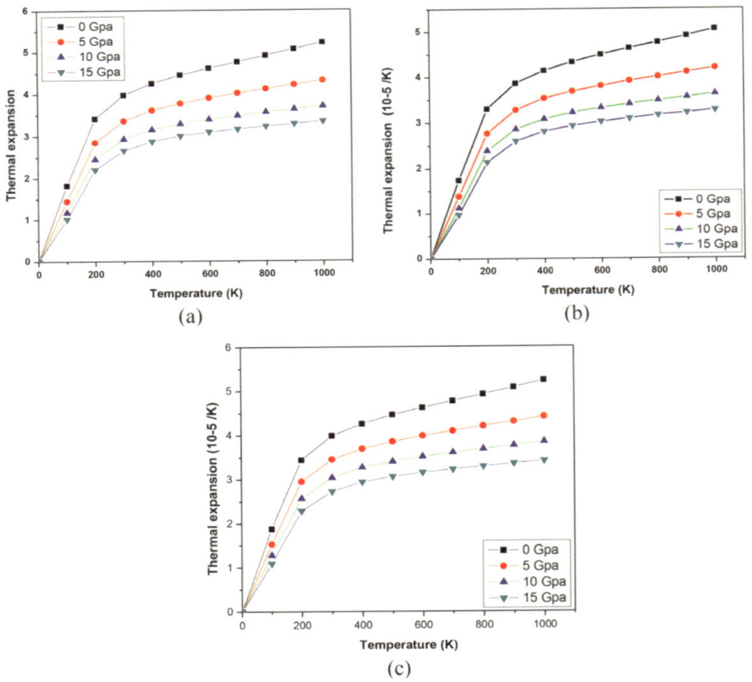

Figure 9.
Dependence of thermal expansion (α) on temperature and pressures for (a) Ba_2MgOsO_6 (b) Ba_2ZnOsO_6 (c) Ba_2CdOsO_6.

pressure at different temperature values. The reason for this decrease of bulk modulus with temperature is that temperature reduces the hardness of a material, while pressure tends to increase the same.

Figure 8(a–c) depicts the variation of specific heat at constant volume (C_V) with temperature and pressure. One can have a clear understanding from the **Figure 8(a–c)** that the escalation of C_V is rapid under the lower temperature values of 0–300 K, but above 300 K a lethargic increase in C_V can be seen, which further becomes constant at high temperature at about 800 K beyond which, it follows the famous Dulong-Petit limit [67]. This variation of C_V for solids is a common observation. The calculated value of C_V at 300 K and 0 GPa of pressure for Ba_2XOsO_6(X = Mg, Zn, Cd) was found to be 223.89, 223.16, 225.04 J mol^{-1} K respectively.

Figure 9(a–c) shows the pressure and temperature dependence of thermal expansion coefficient, 'α' respectively for Ba_2XOsO_6(X = Mg, Zn, Cd). It is clear from the figure that the value of 'α' increase with increasing temperature, the increase in 'α' is found to be rapid under low temperatures values and under higher temperatures values a sluggish increase in 'α' is observed. The main reason for the sluggish increase of 'α' under high temperature values may be the saturation of 'α' beyond 300 K. Pressure has a reverse effect on 'α', increasing pressure decreases the 'α'. Under high pressure values 'α' falls rapidly, same as the increase is observed under low temperatures.

Grüneisen parameter (γ) describes the variation in vibrational frequency of a lattice under the influence of temperature and pressure [68]. Pressure and temperature variation of (γ) for Ba_2XOsO_6(X = Mg, Zn, Cd) is plotted in **Figure 10(a–c)**. The value of (γ) increases with increasing temperature and under pressure a reverse is observed, pressure decreases the value of (γ) and has a lowest value at 15 GPa of

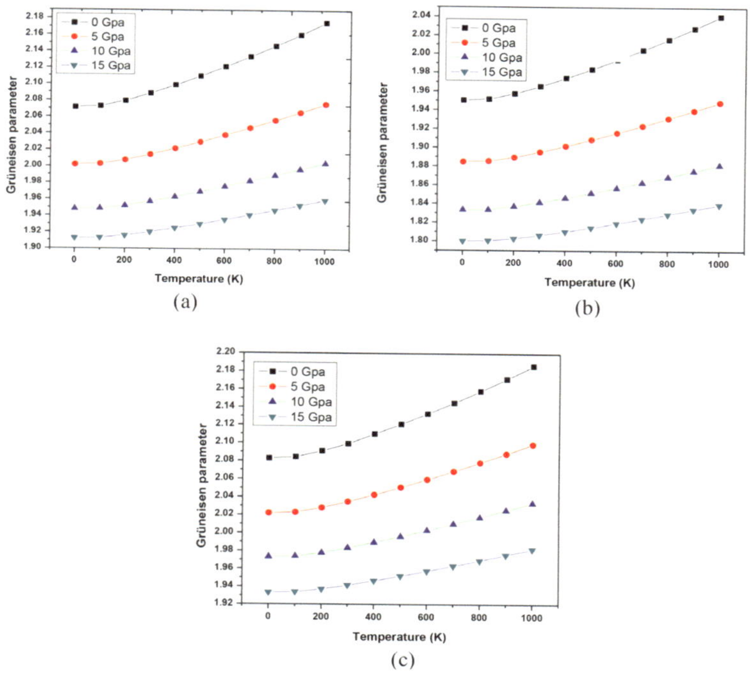

Figure 10.
Dependence of Grüneisen parameter (γ) on temperature and pressures for (a) Ba_2MgOsO_6 (b) Ba_2ZnOsO_6 (c) Ba_2CdOsO_6.

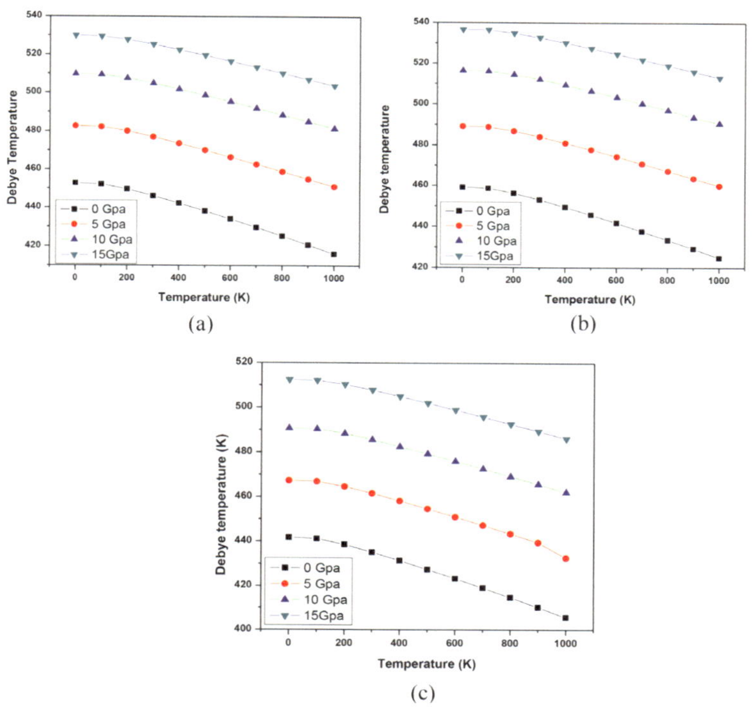

Figure 11.
Dependence of Debye temperature (θ_D) on temperature and pressures for (a) Ba_2MgOsO_6 (b) Ba_2ZnOsO_6 (c) Ba_2CdOsO_6.

pressure. The predicted value of (γ) at 300 K and 0 GPa of pressure is 2.088, 1.965, 2.099 respectively for Ba_2XOsO_6(X = Mg, Zn, Cd).

Debye temperature (θ_D) one of the most important thermodynamic parameter helps to exposes accurate presentation of material properties like specific heat capacity and thermal expansion and also provides the decent understanding about the features of a material under the influence of temperature and pressure. The Debye temperature variation with respect to temperature/pressure is presented in **Figure 11(a–c)** . From these figures it is clear that Debye temperature shows a decreasing trend with increasing temperature and an increasing trend with increasing pressure. The calculated value of Debye temperature at 300 K and 0 GPa is 446.08, 452.97, 435.12 GPa respectively for Ba_2XOsO_6(X = Mg, Zn, Cd).Some part of this work has been recently reported [53].

4. Conclusions

Ab initio calculations on electronic structure, magnetic, elastic, mechanical and thermodynamic properties of cubic double perovskite oxides Ba_2XOsO_6 (X = Mg, Zn, Cd) Ba_2ZnOsO_6 have been reported within density functional theory via full potential linearized augmented plane wave (FP-LAPW) method. The structural investigation reveals the ferromagnetic phase stability for these compounds. The spin polarized electronic and magnetic properties were calculated within generalized gradient approximation (GGA), Hubbard approximation (GGA + U) and mBJ (modified Becke-Johnson approximation). The electronic profile establishes half-metallic nature for these compounds and hence can strength the modern technological domain in terms of spintronic materials. The calculated total spin magnetic moment was found equal to 2 μ_B for all the three compounds. Thus these materials are also looked for magnetic materials. The elastic constants have been calculated and used to predict mechanical stuffs like Shear modulus (G) Poisson ratio (ν) and anisotropic factor (A). The calculated B/G and Cauchy pressure (C_{12}–C_{44}) both characterize the material as brittle. The thermodynamic parameters like heat capacity and Debye temperature have also been predicted in the temperature range of 0–1000 K.

Author details

Sajad Ahmad Dar
Department of Physics, Govt. Motilal Vigyan Mahavidyalya College, Bhopal, Madhya Pradesh, India

*Address all correspondence to: sajad54453@gmail.com

References

[1] Pilania G, Balachandran PV, Kim C, Lookman T. Finding new perovskite halides via machine learning. Frontiers in Materials. 2016;**3**:13285

[2] Moore EA. Computational modelling of inorganic solids. Annual Reports on the Progress of Chemistry, Section A: Inorganic Chemistry. 2008;**104**:46

[3] Kobayashi KI, Kimura T, Sawada H, Terakura K, Tokura Y. Room-temperature magnetoresistance in an oxide material with an ordered double-perovskite structure. Nature. 1998; **395**:677

[4] Iliev MN, Padhan P, Gupta A. Temperature-dependent Raman study of multiferroic Bi_2NiMnO_6 thin films. Physical Review B. 2008;**77**:172303

[5] Fiebig M, Lottermoser T, Frohlich D, Goltsev AV, Pisarev RV. Observation of coupled magnetic and electric domains. Nature. 2002;**419**:818

[6] Das H, De Raychaudhury M, Saha-Dasgupta T. Moderate to large magneto-optical signals in high Tc double perovskites. Applied Physics Letters. 2008;**92**:201912

[7] Krockenberger Y, Mogare K, Reehuis M, Tovar M, Jansen M, Vaitheeswaran G, et al. Sr_2CrOsO_6: End point of a spin-polarized metal-insulator transition by 5d band filling. Physical Review B. 2007;**75**:020404

[8] Feng HL, Arai M, Matsushita Y, Tsujimoto Y, Guo YF, Sathish CI, et al. High-temperature ferrimagnetism driven by lattice distortion in double perovskite Ca_2FeOsO_6. Journal of the American Chemical Society. 2014;**136**: 3235

[9] Stitzer KE, Smith MD, Loye HCZ. Crystal growth of Ba_2MOsO_6 (M = Li, Na) from reactive hydroxide fluxes. Solid State Sciences. 2002;**4**:311

[10] Aharen T, Greedan JE, Bridges CA, Aczel AA, Rodriguez J, MacDougall G, et al. Magnetic properties of the geometrically frustrated S = 1/2 antiferromagnets, La_2LiMoO_6 and Ba_2YMoO_6, with the B-site ordered double perovskite structure. evidence for a collective spin singlet ground state. Physical Review B. 2010;**81**:224409

[11] de Vries MA, Mclaughlin AC, Bos J-WG. Valence bond glass on an fcc lattice in the double perovskite Ba_2YMoO_6. Physical Review Letters. 2010;**104**:177202

[12] Carlo JP, Clancy JP, Aharen T, Yamani Z, Ruff JPC, Wagman JJ, et al. Triplet and in-gap magnetic states in the ground state of the quantum frustrated fcc antiferromagnet Ba_2YMoO_6. Physical Review B. 2011;**84**:100404

[13] Wiebe CR, Greedan JE, Luke GM, Gardner JS. Spin-glass behavior in the S=1/2 fcc ordered perovskite Sr_2CaReO_6. Physical Review B. 2002;**65**:144413

[14] Morrow R, Freeland JW, Woodward PM. Probing the links between structure and magnetism in $Sr_{(2-x)}Ca_{(x)}FeOsO_6$ double perovskites. Inorganic Chemistry. 2014;**53**:7983

[15] Paul AK, Reehuis M, Ksenofontov V, Yan B, Hoser A, Tobbens DM, et al. Lattice instability and competing spin structures in the double perovskite insulator Sr_2FeOsO_6. Physical Review Letters. 2013;**111**: 167205

[16] Kanungo S, Yan B, Jansen M, Felser C. Ab-initio study of low-temperature magnetic properties of double perovskite Sr_2FeOsO_6. Physical Review B. 2014;**89**:214414

[17] Paul AK, Jansen M, Yan BH, Felser C, Reehuis M, Abdala PM. Synthesis, crystal structure, and physical properties of Sr_2FeOsO_6. Inorganic Chemistry. 2013;**52**:6713

[18] Ghimire MP, Hu X. Half metal transition driven by doping effects in osmium double perovskite. arXiv preprintarXiv. 2014;**1408**:1771

[19] Wang H, Zhu S, Ou X, Nu H. Ferrimagnetism in the double perovskite Ca_2FeOsO_6: A density functional study. Physical Review B. 2014;**90**:054406

[20] Dar SA, Srivastava V, Sakalle UK, Pagare G. Insight into structural, electronic, magnetic, mechanical, and thermodynamic properties of actinide perovskite $BaPuO_3$. Journal of Superconductivity and Novel Magnetism. 2018;**31**(10):3201

[21] Dar SA, Srivastava V, Sakalle UK, Pagare G. First-principles investigation on electronic structure, magnetic, mechanical and thermodynamic properties of $SrPuO_3$ perovskite oxide. Materials Research Express. 2018;**5**: 026106

[22] Dar SA, Srivastava V, Sakalle UK, Khandy SA, Gupta DC. A DFT study on structural, electronic mechanical and thermodynamic properties of 5f-electron system $BaAmO_3$. Journal of Superconductivity and Novel Magnetism. 2018;**31**(01):141

[23] Dar SA, Srivastava V, Sakalle UK. A First-principles calculation on structural, electronic, magnetic, mechanical, and thermodynamic properties of $SrAmO_3$. Journal of Superconductivity and Novel Magnetism. 2017;**30**(11):3055

[24] Dar SA, Srivastava V, Sakalle UK, Parey V, Pagare G. DFT investigation on electronic, magnetic, mechanical and thermodynamic properties under pressure of some $EuMO_3$ (M = Ga, In) perovskites. Materials Research Express. 2017;**4**:106104

[25] Dar SA, Srivastava V, Sakalle UK. A combined DFT and post-DFT investigation on cubic $XMoO_3$ (X = Sr, Ba) perovskite oxides. Materials Research Express. 2017;**4**:086304

[26] Zhao S, Wei Z, Dar SA. Insight into the Structural, Electronic, Elastic, Mechanical, and Thermodynamic Properties of $XReO_3$ (X = Rb, Cs, Tl) Perovskite Oxides: A DFT Study. Zeitschrift für Naturforschung A. 2019. DOI: 10.1515/zna-2019-0019

[27] Dar SA, Srivastava V, Sakalle UK. High pressure and high temperature investigation of metallic perovskite $SnTaO_3$. Journal of Molecular Modeling. 2018;**24**:52

[28] Dar SA, Srivastava V, Sakalle UK. Ab initio high pressure and temperature investigation on cubic $PbMoO_3$ perovskite. Journal of Electronic Materials. 2017;**46**(12):6870

[29] Cao K, Li H, Liu S, Cui J, Shen Y, Wang M. MAPbI3-xBrx mixed halide perovskites for fully printable mesoscopic solar cells with enhanced efficiency and less hysteresis. Nanoscale. 2016;**8**:8839-8846

[30] Chen L-C, Weng C-Y. Optoelectronic Properties of MAPbI3 Perovskite/Titanium Dioxide Heterostructures on Porous Silicon Substrates for Cyan Sensor Applications. Nanoscale Research Letters. 2015;**10**:404

[31] Colella S, Mosconi E, Fedeli P, Listorti A, Gazza F, Orlandi F, et al. MAPbI3-xClx mixed halide perovskite for hybrid solar cells: The role of chloride as dopant on the transport and structural properties. Chemistry of Materials. 2013;**25**:4613-4618

[32] Ding J, Zhao Y, Du S, Sun Y, Cui H, Zhan X, et al. Polarization-dependent optoelectronic performances in hybrid halide perovskite MAPbX3 (X = Br, Cl) single-crystal photodetectors. Journal of Materials Science. 2017;**52**:7907-7916

[33] Najeeb MA, Ahmad Z, Shakoor RA, Alashraf A, Bhadra J, Thani NA, et al. Growth of MAPbBr3 perovskite crystals and its interfacial properties with Al and Ag contacts for perovskite solar cells. Optical Materials. 2017;**73**:50

[34] Lu H, Zhang H, Yuan S, Wang J, Zhan Y, Zheng L. An optical dynamic study of MAPbBr3 single crystals passivated with MAPbCl3/I3-MAPbBr3 heterojunctions. Physical Chemistry Chemical Physics. 2016:1-3

[35] Chen L-C, Tseng Z-L, Huang J-K. A Study of inverted-type perovskite solar cells with various composition ratios of (FAPbI3)1-x(MAPbBr3)x. Nanomaterials. 2016;**6**:183

[36] Yuan S, Wang J, Yang K, Wang P, Zhang X, Zhan Y, et al. High efficiency MAPbI3-xClx perovskite solar cell via interfacial passivation. Nanoscale. 2018; **10**:18909-18914

[37] Marjerrison CA, Thompson CM, Sharma AZ, Hallas AM, Wilson MN, Munsie TJS, et al. Magnetic ground states in the three Os_6+ ($5d_2$) double perovskites Ba_2MOsO_6 (M=Mg, Zn, and Cd) from Néel order to its suppression. Physical Review B. 2016;**94**:134429

[38] Dar SA, Srivastava V, Sakalle UK, Pagare G. Insight into electronic structure, magnetic, mechanical and thermodynamic properties of double perovskite Ba2MgReO6: A first-principles investigation. Computational Condensed Matter. 2018;**14**:137-143

[39] Merabet B, Alamri H, Djermouni M, Zaoui A, Kacimi S, Boukortt A, et al. Optimal bandgap of double perovskite La-substituted Bi_2FeCrO_6 for Solar Cells: An ab initio GGA+U Study. Chinese Physics Letters. 2017;**34**(1):016101

[40] Dar SA, Srivastava V, Sakalle UK. Structural, elastic, mechanical, electronic, magnetic, thermoelectric and thermodynamic investigation of half metallic double perovskite oxide Sr_2MnTaO_6. Journal of Magnetism and Magnetic Materials. 2019;**484**:298-306

[41] Sahnoun O, Bouhani-Benziane H, Sahnoun M, Driz M. Magnetic and thermoelectric properties of ordered double perovskite Ba_2FeMoO_6. Journal of Alloys and Compounds. 2017;**714**:704

[42] Dar SA, Sharma R, Srivastava V, Sakalle UK. Investigation on the electronic structure, optical, elastic, mechanical, thermodynamic and thermoelectric properties of wide band gap semiconductor double perovskite Ba_2InTaO_6. RSC Advances. 2019;**9**:9522

[43] Dar SA, Srivastava V, Sakalle UK. Ab-initio DFT based investigation of double perovskite oxide Ba_2CdOsO_6 with cubic structure. Computational Condensed Matter. 2019;**18**:e00351

[44] Thompson CM, Marjerrison CA, Sharma AZ, Wiebe CR, Maharaj DD, Sala G, et al. Frustrated magnetism in the double perovskite La_2LiOsO_6: A comparison with La_2LiRuO_6. Physical Review B. 2016;**93**:014431

[45] Dar SA, Srivastava V, Sakalle UK, Parey V, Pagare G. A combined DFT, DFT+U and mBJ investigation on electronic structure, magnetic, mechanical and thermodynamics of double perovskite Ba_2ZnOsO_6. Materials Science and Engineering B. 2018; **236-237**:217-224

[46] Faizan M, Khan SH, Murtaza G, Khan A, Khenata R, Mahmood A, et al. Structural, elastic, electronic and magnetic properties of Ba_2XOsO_6 (X = Li, Na, Ca) double perovskites. Indian Journal of Physics. 2016;**90**:1225

[47] Dar SA, Srivastava V, Sakalle UK, Parey V. Electronic structure, magnetic, mechanical and thermo-physical behavior of double perovskite Ba_2MgOsO_6. European Physical Journal Plus. 2018;**133**:64

[48] Otero-de-la-Roza A, Abbasi-Perez D, Luaea V. Gibbs2: A new version of the quasiharmonic model code. II. Models for solid-state thermodynamics, features and implementation. Computer Physics Communications. 2011;**182**:2232

[49] Otero-de-la-Roza A, Luaea V. Equations of state and thermodynamics of solids using empirical corrections in the quasiharmonic approximation. Physical Review B. 2011;**84**:184103

[50] Wu Z, Cohen RE. More accurate generalized gradient approximation for solids. Physical Review B. 2006;**73**:235116

[51] Blaha P, Schwarz K, Madsen GKH, Kuasnicke D, Luitz J. Introduction to WIEN2K, an Augmented Plane Wane plus Local Orbitals Program for Calculating Crystal Properties. Vienna, Austria: Vienna University of Technology; 2001

[52] Perdew JP, Burke K, Ernzerhof M. Generalized gradient approximation made simple. Physical Review Letters. 1996;**77**:3865

[53] Tran F, Blaha P. Accurate band gaps of semiconductors and insulators with a semilocal exchange-correlation potential. Physical Review Letters. 2009;**102**:226401

[54] Petukhov AG, Mazin II. Correlated metals and the LDA+U method. Physical Review B. 2003;**67**:153106

[55] Novak P, Kunes J, Chaput L, Pickett WE. Exact exchange for correlated electrons. Physica Status Solidi B: Basic Solid State Physics. 2006;**243**:563

[56] Aisimov VI, Solovye IV, Korotin MA, Czyzyk MT, Sawatzky GA. Density-functional theory and NiO photoemission spectra. Physical Review B. 1993;**48**:16929

[57] Hou YS, Xiang HJ, Gong XG. Lattice-distortion induced magnetic transition from low-temperature antiferromagnetism to high-temperature ferrimagnetism in double perovskites A_2FeOsO_6 (A = Ca, Sr). Scientific Reports. 2015;**5**:13159

[58] Monkhorst HJ, Pack JD. Special points for Brillouin-zone integrations. Physical Review B. 1976;**13**:5188

[59] Charpin T. A Package for Calculating Elastic Tensors of Cubic Phases Using WIEN. Paris, France: Laboratory of Geometrix F-75252; 2001

[60] Birch F. The effect of pressure upon the elastic parameters of isotropic solids, according to Murnaghan's theory of finite strain. Journal of Applied Physics. 1938;**9**:279

[61] Dimitrovska S, Aleksovska S, Kuzmanovski I. Prediction of the unit cell edge length of cubic A_2^{2+} BB'O6 perovskites by multiple linear regression and artificial neural networks. Central European Journal of Chemistry. 2005; **3**(1):198

[62] Mehl MJ, Klein BK, Papaconstantopoulos DA. (Intermetallic compounds); principle and practice. In: Westbrook JH, Fleischeir RL, editors. Principles. Vol. 1. John Wiley and Sons; 1995

[63] Voigt W. Lehrbush der Kristallphysik. Leipzig: Taubner; 1928

[64] Schreiber E, Anderson OL, Soga N. Elastic Constants and Measurements. New York: M.C Graw Hill; 1973

[65] Hill R. The elastic behaviour of a crystalline aggregate. Proceedings of the Physical Society (London). 1952;**65**:349

[66] Pugh SF. Relations between the elastic moduli and the plastic properties of polycrystalline pure metals. Philosophical Magazine. 1954;**45**:823

[67] Petit AT, Dulong PL. Sur quelques points importants de la théorie de la chaleur. Annales de Chimie Physique. 1819;**10**:395

[68] Quiang L, Duo-Hui H, Qi-Long C, Fan-Hou W. Phase transition and thermodynamic properties of $BiFeO_3$ from first-principles calculations. Chinese Physics B. 2013;**22**:037101